INVASIVE
PLANT
MEDICINE

"Timothy Lee Scott shows how wrongheaded it is to single out other species as harmful. In nearly every case, the blame for damage done by so-called invasive species lies with us, when we have created an imbalance that opens opportunities for new species to move in. Tim goes beyond simply removing blame from our fellow species. He shows how erstwhile invaders can teach us how to heal damaged ecosystems and ourselves."

TOBY HEMENWAY, AUTHOR OF *GAIA'S GARDEN:*
A GUIDE TO HOME-SCALE PERMACULTURE

"This is an important, insightful book that should be read by all involved in herbal medicine or plant conservation, and more importantly, for all of us who should be questioning authority. In these times of rapid change it is refreshing to find such competent questioning of 'established truths.' Thanks, Tim!"

DAVID HOFFMAN, BSc, FNIMH,
MEDICAL HERBALIST AND AUTHOR OF
MEDICAL HERBALISM AND *HERBAL PRESCRIPTIONS AFTER 50*

"So, be warned, this is a dangerous book. Tim Scott will change how you see 'invasives,' will make you question what you have been taught about them, force you to reexamine what you have read in newspapers, and demand you look more closely at what 'experts' have said."

STEPHEN HARROD BUHNER, AUTHOR OF
THE SECRET TEACHINGS OF PLANTS

INVASIVE PLANT MEDICINE

The Ecological Benefits
and Healing Abilities
of Invasives

Timothy Lee Scott

Healing Arts Press
Rochester, Vermont • Toronto, Canada

Healing Arts Press
One Park Street
Rochester, Vermont 05767
www.HealingArtsPress.com

Healing Arts Press is a division of Inner Traditions International

Text paper is SFI certified

Note to the reader: *This book is intended as an informational guide. The remedies,
approaches, and techniques described herein are meant to supplement, and not to be a
substitute for, professional medical care or treatment. They should not be used to treat a serious
ailment without prior consultation with a qualified health care professional.*

Library of Congress Cataloging-in-Publication Data
Scott, Timothy Lee.
  Invasive plant medicine : the ecological benefits and healing abilities of invasives /
Timothy Lee Scott.
    p. cm.
  Includes bibliographical references and index.
  Summary: "The first book to show how plants wrongly identified as harming the
environment actually restore the earth's ecosystem and possess powerful healing
properties"—Provided by publisher.
  ISBN 978-1-59477-305-1
  1. Invasive plants—Ecology. 2. Invasive plants—Therapeutic use. 3. Materia medica,
Vegetable. I. Title.
  SB613.5.S36 2010
  581.6'2—dc22
                                                                    2010014758

Printed and bound in the United States by Lake Book Manufacturing
The text paper is 100% SFI certified. The Sustainable Forestry Initiative® program
promotes sustainable forest management.

10   9   8   7   6   5   4   3   2   1

Text design and layout by Virginia Scott Bowman
This book was typeset in Garamond Premier Pro and Gill Sans with Baskerville and Gill
Sans as display typefaces

To send correspondence to the author of this book, mail a first-class letter to the author
c/o Inner Traditions • Bear & Company, One Park Street, Rochester, VT 05767, and we
will forward the communication.

Vegetus Vigoratus Veritas

To All the Lively Ones Who Speak the Truth,
Yet Continue to be Trampled Upon

The Piously Prolific Pioneering Plants

Contents

PART 3

Guide to Invasive Plants: Medicine and Ecological Roles

Foreword

Analogy can connect body and mind, objective space and subjective space, and the animal, plant, and mineral realms in a way logic cannot. It is the key to the groundbreaking correlations Le Brun makes between the environmental degradation of our physical world and the ravages suffered by the imaginal realm of our minds. The relationship between the disappearance of the great mammals and the blue whale and the great rebels of times past is the same insidious and pervasive decay as the depreciation and adulteration of language and the genetic modification of the foods we eat. In all the cases she cites, it is clear that the shackles placed on the human imagination have made possible the environmental and social degradation that pervades our modern world. For Annie Le Brun, the most horrendous ecological catastrophe is found in "the growing impossibility to imagine the symbolic exchange that never stops occurring between ideas, beings, and things."

JON GRAHAM, TRANSLATOR'S FOREWORD,
*THE REALITY OVERLOAD: THE MODERN WORLD'S
ASSAULT ON THE IMAGINAL REALM*

CENTRAL TO OUR TIME is the growing recognition that the linear mind, what Robert Bly sometimes refers to as the statistical mentality, has reached the limits of its perceptual thought. This is especially true when it comes to understanding the subtle and all-pervasive interdependencies, communications, and layered subtleties of the Natural world. In other words, applying *A* to *B* to *C* thinking to our world doesn't work, in fact, the great preponderance of evidence is showing that not only does it not work, it is also the source of a great many of our ecological problems.

The most important things in the ecological functioning of this planet, it turns out, tend to be invisible, that is, not readily perceivable, to linear thinking. Our inability to work with these crucial invisibles can be traced to a problem in the assumptions that underlie the type of thinking that most of us in the West are trained in from the age of six onward.

Unfortunately, those on the left are as equally guilty of this as those on the right; the problems facing us will not be solved by simplistic social injunctions such as recycling or the use of compact fluorescent lightbulbs. Those types of solutions come out of the same kind of linear thinking that got us into this mess. It just uses a different power dynamic than is used on the right: a social shaming that later is translated into heavy-handed legislation that, it is insisted, will solve our problems if only everyone is forced to comply with it. That approach is simply too superficial, too unaware of the generative causes of most of the problems, and too generative itself of unexamined consequences to be of use. It allows the majority of people to engage in simple behaviors that, ultimately, support them escaping the much more difficult, and necessary, task of deeply examining the complexity of our ecological problems.

Nor will the application of typical conservative or right-leaning thinking solve those problems, that is, the use of force against the natural world—specifically the imposition of scientific rationalism expressed through corporate technology in an attempt to force Nature to bend to our will, to allow us to continue ever more to exceed the carrying capacity of our planet.

What is necessary is to begin to think outside the boxes that have been placed around our perceptions, to literally begin to see what is right before us, and then to act on our own self-intuited wisdom and perceptions to address the problems that each of us, individually, perceives as necessary to address. The problems that face us are simply too complex for top-down solutions to work. They need to be found by people on the ground, people inhabiting their own unique ecosystems and who understand from that embedding just what those ecosystems need. The solutions must come out of the individual genius of millions, not the few. And these solutions cannot be behavioral solutions imposed through top-down legislation. Totalitarianism under the guise of saving the planet is still totalitarianism. Worse, the evidence is abundantly clear: that kind of approach will create more problems than it will solve. It is based on flawed assumptions.

The main problem facing us is not so much our behaviors but how we think, for inside that thinking, just as the oak and the shade are inside the acorn, are the behaviors that lead to the ecological damage of our world. Merely altering behaviors without addressing the underlying thinking that gives rise to them is only to substitute one set of problems for another.

Thinkers such as the great French poet and writer Annie Le Brun with her book *The Reality Overload* and the historian John Ralston Saul with his book *Voltaire's Bastards: The Dictatorship of Reason in the West* have joined such writers as Gregory Bateson (who foresaw these problems in the 1960s) and Aldo Leopold (who foresaw them even earlier) in exploring, in depth, the truth that the source of the ecological damage to this planet lies in our thinking. The left has had some intimation of this for some time, but they have used the same flawed approaches as the right in their attempts to address it. They have confused thinking with a lack of information about the planet and the impact of industrial society upon it. The problem is not lack of information and, truthfully, if you look at what most of humanity has known about this planet for many millennia, it never has been. The problem lies in the way we think, not what we think.

Gregory Bateson was reaching toward this when he said that the dynamics of our thought since Descartes represented an epistemological mistake. He believed that that mistake would prove to be more serious than, as he put it, "all the minor insanities that characterized those older epistemologies which agreed upon fundamental unity." We have lost, he said, a "sense of aesthetic unity" and that loss underlies many, if not most, of the problems we now face.

The sense of aesthetic unity is something that is felt rather than thought. And it is this very same thing that Annie Le Brun is discussing when she talks of the damage done us in the imaginal realm. And it is that very same thing that I am working with when I say that Descartes dictum I think therefore I am is and was so dangerous to our habitation of this planet specifically because it affirms its opposite: If you do not think you are not. Thus the level of thinking of any organism can henceforth be used to determine its value. But more problematical is that Descartes' dictum also means If you feel you are not and that removal of value from feeling and from those who do feel in and of itself makes it nearly impossible for succeeding generations to engage the imaginal realm, to experience the sense of aesthetic unity that comes from it, to escape the dictatorship of reason.

It makes it very difficult to feel caring for Nature if the feeling sense itself is suspect. But the problem goes deeper than that. For it is our capacity to feel, to engage the imaginal realm, to daily experience a sense of aesthetic unity that allows human beings to perceive the invisibles that surround us every minute of our lives and which are the most important elements of ecosystem health and function. Without the feeling sense, we cannot even perceive the subtle interactions that occur in ecosystems, most of which are invisible to the rational mind. And to support the restoration of the ecosystems of the world we must be able to perceive these subtle interactions, understand their nature, what they do, and just why they do it. We have to begin to work at that level of subtlety ourselves if we truly are to become ecologists in anything but name. And that subtlety is nowhere more present than in plants and plant communities.

Plants, the subject of Tim Scott's book, are, perhaps, one of the most important ecosystem stabilizers that exist on this planet. It is they who actively work to keep oxygen content within the parameters that allow mammalian life on this planet to exist, they who remove and work to inactivate pollutants in the soil and water, they who are actively involved in the reduction of greenhouse gases, and they who are the greatest chemists on this planet. Their chemical production is exceptionally sophisticated, each plant making between one hundred and one thousand different so-called secondary chemicals, each of which is created and released into the ecosystem for a reason. Most of them are part of a complex chemical signaling and communication network and maintenance system designed to maintain the homeodynamis of the planet and any ecosystem in which they grow. They are, in essence, expressed out of the Gaian planetary system itself to fulfill specific ecosystem functions, primary among which is that maintenance of homeodynamis.

So, the most important thing to do, ecologically, whenever you encounter a plant, is to ask yourself what that plant is doing. What Gaian dynamics and purposes are in play? We have to approach this question with great humility. We can't presume to know the answer before we start. We have to, first, learn to see what is right before us and that means getting rid of a lot of our programming. Then we must let the plant and ecosystem themselves teach us about what is occurring.

We need to understand that Nature doesn't make mistakes, that Earth is, at minimum, 3.5 billions years old, and that Earth has been engaging in this process a lot longer than our species has existed. We have to understand that what we are looking at predates the human, that Gaian timelines are much longer than ours. We need to understand that processes that no scientists understand are occurring on both very large and very small scales. We have to step outside the human paradigm if we are to understand what is occurring with the appearance and behavior of any plant we encounter. So, when we see "invasive" plants moving wholesale into new ecosystems, we need to ask, in all humility, "What are they doing? What is their purpose?"

The book that you now hold in your hands is part of a counter movement that holds at its core that very question. Unlike too many other books, it actually struggles with the very difficult process of finding an answer. It is a crucial and necessary look at the importance of invasive plants. These plants, it turns out, play crucial parts in the restoration of our ecosystems; they are expressions of Gaia, sent to work in specific ways in specific places that need what they uniquely can do. They are, rather than destructive pests, ecological interventions generated out of the vast, long-scale movements of the Earth, intended to solve specific ecological problems.

So, be warned, this is a dangerous book. Tim Scott will change how you see "invasives," will make you question what you have been taught about them, force you to reexamine what you have read in newspapers, and demand you look more closely at what "experts" have said. In here you will find how fear of the other has been projected from the human realm onto the plant world, see connections between how we treat immigrants and how we treat immigrant plants, begin to understand the fear that is underneath current drives for purity, of body, culture, and ecosystems.

"Invasive" plants are messengers you see, and as Tim notes in the book, "So many times, the messenger has been killed."

STEPHEN HARROD BUHNER

Stephen Harrod Buhner is an Earth poet and the award-winning author of fourteen books on nature, indigenous cultures, the environment, and herbal medicine, including *The Secret Teachings of Plants, The Lost Language of Plants,* and *Sacred Plant Medicine.* He comes from a long line of healers including Leroy Burney, Surgeon General of the United States under Eisenhower and Kennedy, and Elizabeth Lusterheide, a midwife and herbalist who worked in rural Indiana in the early nineteenth century. The greatest influence on his work, however, has been his great-grandfather C. G. Harrod, who primarily used botanical medicines, also in rural Indiana, when he began his work as a physician in 1911. Stephen lectures throughout the United States on herbal medicine, the sacredness of plants, and the intelligence of Nature.

Acknowledgments

I AM GRATEFUL FOR all of the help I had along the way that has made this book possible. Many people have assisted this work, some through their mere presence, others through their feedback, some through their writings, and others through fieldwork and study. I give thanks for all of those who have come before in service of the natural world and who have passed down knowledge about these plants for thousands of years—the poets, healers, gardeners, and visionaries. I give thanks to the many voices who have spoken for invasives and dared tread the treacherous path of standing up on behalf of these mistaken plants. I am indebted to those who took time out of their busy lives to help color the plants with valuable knowledge and firsthand experience, and I know I have missed many, and I apologize that I cannot thank everyone individually.

First and foremost, I am grateful for my wife, Colleen, and her endearing presence and enduring patience as I embarked on this newfound journey into insanity called writing a book. For my children, Osha and Rowan, who continually remind me of what's most important in life. Also, I am thankful for my parents and their constant support in my life choices and for letting me roam wild in the woods.

I give many thanks to my friends Rebecca Sunter and Monique Bonneau for their continued encouragement and critique, from the beginning as an article with a tiny seed of an idea, to the end of a full-length manuscript. Becca's keen eye in reviewing my manuscript tied many loose ends and helped me explore the deeper layers of writing as art, and Moe's wonderful illustrations captured the spirit of the plants through line and color. And much appreciation to my friend and soul sister Julie McIntyre, who provided photos of some of the southwestern plants.

I am grateful to Jon Graham and Inner Traditions • Bear and Company for seeing the potential in this book and setting me forth on this adventure. Many thanks to the remediation expert Paul Schwab for finding the initial phytoremediation studies that confirmed my beliefs of the ecological benefits of invasives and pointed me in the right direction. I am thankful for all of the elders in the American herbal renaissance that have tied together the herbal traditions from all over the world, held the spirit of the plant medicines intact, and helped show the value of and compassion for these widespread plants. I appreciate the generosity of Todd Hardie and his sharing of his love of the bees and of purple loosestrife, and for Steven Foster, who rounded out the photos I needed with his superb plant eye. I thank Dale Pendell for his high plant poems and green-wise thinking that takes me on trips each time I read them. And I'm especially honored by Rosemary Gladstar taking time to share precious jewels of herbal wisdom for this book.

Finally, deep gratitude and affection goes to Stephen Buhner, for without his work, none of this would have been possible.

Introducing the Weed

Every plant is a teacher—
but as in every crowd,
there are always
a few loudmouths.

DALE PENDELL,
LIVING WITH BARBARIANS

THERE WAS A TIME, now long forgotten, around the advent of agricultural civilizations some ten thousand years ago, that humans began to look at plants differently. Before this, all was a wild garden with diverse flora and fauna, and each of these living beings had a place. But with the invention of the crop, people began to discriminate between the different plants and chose to keep some and remove others. A plant that did not serve human needs or that interfered with the crops was deemed a *weed*. This marked a shift in the paradigm of paradise, and humans began severing themselves from nature in a paramount way. The desire and attempt to keep the wild at bay was passed down through the generations, and such thinking is predominant to this day.

The nature of a weed is opportunistic, and we, as humans, have created enormous holes of opportunity for these plants to fill. They have adapted to be at our side, waiting for those favorable times to cover the exposed soils that we continually create. With ever-changing genetics of form, function, and transmutation, weeds have evolved to withstand the punishments that humans unleash upon them.

1

Kudzu, 1945. From H. H. Biswell, USDA-NRCS Plants database.

Weeds are especially adapted to adapt.

For tens of thousands of years, people have transported and intention-ally introduced plants all over the world for food, fiber, medicine, orna-mentation, and scientific curiosity, and because this practice has continued to the present day, we humans have been complicit in and have encouraged the spread of plants. Nowadays, the common plants we see throughout our meadows, countryside, and city streets—such as plantain, mullein, Saint-John's-wort, burdock, chicory, coltsfoot, fennel, and daylily—are alien spe-cies that did not grow here until the first Europeans arrived. Both a Native American and Chinese name for common plantain translate as "white man's footsteps," referring to the fact that this plant followed along the colonizing trail of Europeans. One plantain species has sword-shaped leaves (*lanceolata*) with wound-healing abilities, but instead of complaining about this plant, the indigenous herbalists made good use of it as medicine—for it was needed. I do not know if the first such plant arrivals would have been considered invasive some five hundred years ago, but they certainly were foreign, just like the knotweeds and loosestrifes of today. Over time, though, these plants have found an ecological niche in a dynamic equilibrium among the different species within the landscape.

Within their niche, all plants serve ecological functions for their environment. Mullein, for example, blankets the land where fire has cleared forests. In this, it appears as though the plant is invading the land, but after a year or two, new plant species emerge and diversity expands. Mullein acts as a kind of earth balm that eases and covers with its leaves the internal burns and helps regenerate new growth—which it also happens to do for human lungs.

Forests are the lungs of the earth, you know.

Therefore, all colonizing plants offer medicine; some provide food for human, animal, and other inhabitants, some protect the land after improper clearing and use, some renew degraded soils, some cleanse the waters, and some break down and clean up toxins and pollutants in the soil and air. These plants are here for a reason. They are here to serve essential ecological functions *and* they are here for us to use as medicine.

> *But there are the rampant, freaky plants from faraway lands,*
> *those with loud, annoying voices that scream throughout the*
>   *world.*
> *They stand up and say, "Here I am," with an arrogant smirk,*
> *and they go about spreading the Good Word cheerfully.*
> *People can put up with the voices for only so long.*
> *If they do not dampen their own volume,*
> *human conditioning resorts to force to make them stop.*
> *But what if these voices actually say something important?*
> *Do we stop to listen? Do we know why this strange one has*
>   *come bearing exotic fruits?*
>
> *So many times, the messenger has been killed.*

I consider the land I live on a sanctuary for the plants that live here too. I have established and protected woodland gardens of endangered native species such as ginseng, goldenseal, blue cohosh, bloodroot, and pink lady's-slipper and have for a number of years been a member of United Plant Savers, which is an organization dedicated to protecting North American medicinal plants in danger of extinction. I have the honor of working with the plant world every day. I run an established herbal apothecary and practice as an herbalist, wild-crafting many of my own medicines, gardening, and even wrestling with invasive species on my own land. I feel a deep love for my surrounding landscape, and I am committed to keeping it beautiful.

The spark for this book was ignited many years ago, when my wife first imparted to me the idea that there is no such thing as a weed. From then on, I've tried to follow the assertion of Ralph Waldo Emerson: a weed is "a plant whose virtues have not yet been discovered." I first wrote an article to vent some of my frustrations about this one-sided paradigm. I gnawed and tugged on the article for almost a year, having never before experienced the passion of writing it imparted to me. I began talking about it, and I shared with others what I succumbed to as a finished article. The feedback was encouraging; many wanted to read more. I thought, "Huh . . . this could be a book."

So I sought out others who have done this before. I could find only bits and pieces of my ideas in others' writings—no comparable books focused on the medicinal and ecological benefits of these persistent plants. I arranged a deal with myself: I would send out some book proposals to publishers, and if anyone wanted me to write the book, then I would have to write it. The universe responded with a few nibbles and then a big bite. I had to keep to my word to write this book, and I have set out with high hopes and deep convictions.

I am here in defense of these plants, and I am set to demonstrate that the population of invasives is wrongly convicted. I lay out the evidence that these plants serve essential ecological functions and actually benefit the environment, with prospects to strengthen local economies and with their ways of stimulating our health and healing disease.

I have been fortunate to be exposed to the work of quite a few dedicated people defending pandemic plants. They all provide a piece of the puzzle and have varying takes on invasives, setting the foundation for this work. I am comforted by the wild and raucous people who defend invasives: the ecologists, herbalists, journalists, farmers, and gardeners who take a stand by questioning the science, politics, psychology, and ecology of invasive plants. The herbal medicine world has the greatest camaraderie with these widespread plants, with the strong healing remedies they provide, and some individuals are pressing buttons in the permaculture discussions by offering alternative perspectives of exotic species, where native plants are considered of prime importance. We, who speak of these plants without malice, are like weeds and outcasts in a way, living on the outskirts of civilized thought and expressing a multitude of voices that represent the wild fluctuations of Nature. Yet no one has put all of these pieces together to create a deep, multifaceted, ecological view of invasive plants. This book is my best attempt at doing so.

> *All the technical information was stolen from reliable sources and I am happy to stand behind it.*
>
> EDWARD ABBEY

> Noxious weeds. What made so many of us hate these plants so much? Sure, I can understand their annoyance. I myself have done battle against blackberry and poison ivy. Yet when did this aversion to nature taking its due course cut so deep into our collective psyche? Who has propagated the idea of the noxious weed and created this war? And why? I wish to know.

In our modern world, it seems as if all of us (humans, plants, animals, microbes, etc.) are meandering around, looking for a place to settle down—some move here, others there, and a few are unaccommodating to those that are nearby. We all have taken root where conditions are right and have become a new expression of growth, unique in our circumstances of life, with never before the means to attain these potentials. This present time on the good planet Earth marks the birth of something new: a global mosaic of intermingling forms and colors with changes to the land-scape, and a group of plants erasing the scars of the Industrial Age.

Whether we like them here or not.

> *It spread beyond England very speedily. Soon in America, all over the continent of Europe, in Japan, in Australia, at last all over the world, the thing was working towards its appointed end. It was bigness insurgent. In spite of prejudice, in spite of law and regulation, in spite of all that obstinate conservatism that lies at the base of the formal order of mankind, the Food of the Gods, once it had been set going, pursued its subtle and invincible progress.*

H. G. WELLS, *THE FOOD OF THE GODS*

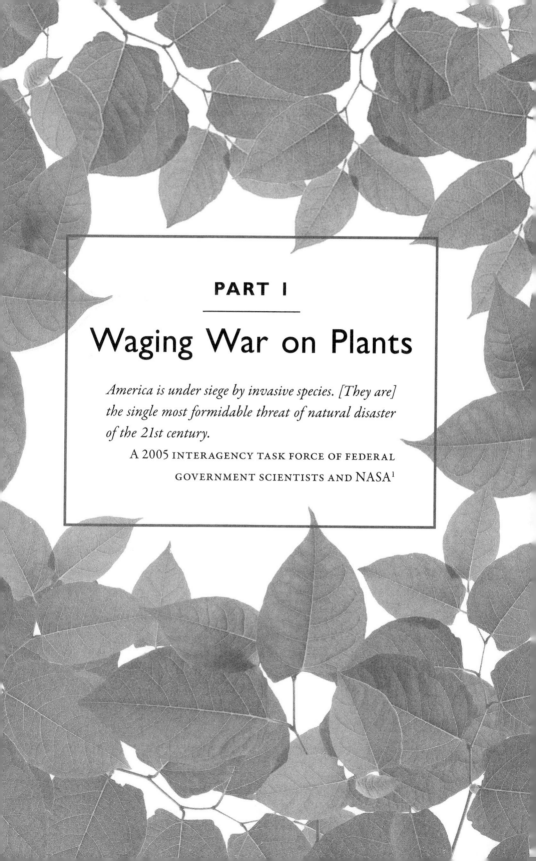

PART I

Waging War on Plants

America is under siege by invasive species. [They are]
the single most formidable threat of natural disaster
of the 21st century.

A 2005 INTERAGENCY TASK FORCE OF FEDERAL
GOVERNMENT SCIENTISTS AND NASA[1]

But while men slept, his enemy came and sowed tares among the wheat, and went his way.

MATTHEW 13:25

THERE IS A WAR being conducted with countless silent casualties. We are told it is better this way: to kill our offenders instead of trying to understand them and listen to their side of the story. We are told we must stop them, for they are destroying and taking over our lives. This war is taking place all over our world, and there is a deadly secret slowly invading our own backyards.

And it is not the plants, you think, that are the problem.

A war—either real *or imagined*—must be waged in order to gather public support and funding from the government. As much hype and vigor as possible is added into the public discourse: *War on Invasive Species, War on Terrorism, War on Drugs, War on Disease, War on Poverty* . . .

It is interesting to note that all our instruments of destruction and war come from the same companies, in the same factories, with the same mentality. The same raw materials are used to create the deadly poisons both for human warfare and for the everyday War on Pests: sarin gas, DDT, Agent Orange, and the rainbow herbicides by the notorious Monsanto, DuPont, and Dow Chemical Company. These war factories are good at only two things: death and destruction.

We must remember, though, that when we fight against Nature, Nature will surely prevail in the end.

1

The Politics
of Prolific Plants

*The greatest service which can be rendered any country is to
add a useful plant to its culture.*

THOMAS JEFFERSON[1]

UNTIL RECENTLY, IT HAS been the practice of the United States
government—with the support of numerous presidents—to promote the
introduction of plants for food and economic, medical, and landscaping
purposes. The founders of this country enthusiastically encouraged a wide
variety of plants. In fact, our sixth president, John Quincy Adams, established a national policy that stated, "The United States should facilitate
the entry of plants of whatever nature whether useful as a food for man or
the domestic animals, or for purposes connected with . . . any of the useful
arts."[2] There once was a United States Office of Plant Introduction, and
it claimed to bring in two hundred thousand nonnative species. The U.S.
Soil Conservation Service promoted now-invasive multiflora rose for erosion control, highway divides, and living fences, and the Department of
Agriculture planted the rampantly widespread kudzu, tamarisk, Russian
olive, and others for fodder, windbreaks, erosion control, and other useful
purposes.

The significant problems we face today cannot be solved at the
same level of thinking we were at when we created them.

ALBERT EINSTEIN

By the dawn of the twentieth century, laws were passed by the U.S. Congress to control the plants that impeded the progress of the Great Agricultural Machine. The regulations started with the Lacey Act of 1900, followed by the Plant Pest Act, the Plant Quarantine Act, and the federal Noxious Weed Act of 1974, in which were targeted plants that "can directly or indirectly injure crops, other useful plants, livestock, poultry or other interests of agriculture, including irrigation, navigation, fish and wildlife resources, or the public health."[3] Farmers were the primary promoters of these first bills that would protect crops and livestock from the wilds of nature. It was not until Executive Order 13112, signed in 1999 by President Clinton, that billion-dollar funds were allocated to promote widespread appeal and greater influence to "rapidly respond" to "an alien species whose introduction does or is likely to cause economic or environmental harm or harm to human health."[4] This has led to a halt on importing virtually any exotic plant that hasn't been cleared as a threat, even though scientists find it impossible to determine which plants might be threats in the first place. No one knows the exact number of exotic species of plants in the United States. The widest range is from thirty thousand to fifty thousand, with maybe five thousand to seven thousand becoming naturalized.[5] According to Dr. Daniel Simberloff, a professor of ecology and evolutionary biology at the University of Tennessee, of the country's seven thousand exotic species that have naturalized (plants, animals, insects, etc.), only seven hundred are considered invasive. The rest of them fill a niche and incorporate themselves into the natural surroundings.[6] This means that with a total of one hundred fifty thousand species living in this country, less than ½ percent of all of them fit into the category that we put so much energy and money into combating.

The National Invasive Species Council (NISC) was formed in accordance with Executive Order 13112 and created oversight to authorize and fund programs geared toward invasive plants, animals, and microorganisms. Council members include three cochairs: the secretaries of the Interior, Agriculture, and Commerce; the secretaries of State, Defense, Homeland Security, Treasury, Transportation, and Health and Human Services; and the Administrators of the Environmental Protection Agency, the U.S. Agency for International Development, and the U.S Trade Representative.

The federal Invasive Species Advisory Committee, along with eight working groups, provides information and advice to the council to set up the National Management Plan for invasive species. Surprisingly—or not—the working group for control and management is headed by a Monsanto employee, and a representative on the committee hails from BASF chemical corporation (companies specializing in the manufacture of herbicides). Seats on the committee also include academic representatives specializing in "bio-based pest management." In addition to this apparent conflict of interest, many state organizations have deep roots within the chemical industry, including the California Exotic Pest and Plant Council that was established by a Monsanto executive.[7]

What are the motives driving these advisories?

These policies based on opinions of good plants versus bad plants have far-reaching effects, and to the fight, the federal government has allocated hundreds of millions to billions of dollars every year since 1999. Residential control pays out at least half a billion dollars a year, and golf courses alone spend one billion dollars per year with some outstanding estimates, such as the Cornell University claim that each year the U.S. spends 34.7 billion dollars in fighting noxious weeds.[8] Whatever the exact number, that's billions of dollars spent every year on so-called invasive plants! These policies divert vast resources that could be better spent on more imperative issues such as habitat preservation, the study of plant medicines, renewable resources, and efforts to repopulate the land with those unique plants that are on the brink of extinction. Yet only one question is considered in these discussions: "How the heck do you get rid of them bad ones over there?"

In the fiscal year 2006, the federal government allocated more than one billion dollars to the Department of Agriculture to fight invasive species, with the majority spent on control efforts (465 million dollars) and rapid response efforts (257 million dollars)—a grand total of 722 million dollars essentially for herbicides and pesticides.

> *How do we measure the long-term effects to the environments*
> *where nonnative plants move in if a rapid response is mandated*
> *and funded by the national council?*

As for the rest of the funds, nearly the same amount granted to control efforts (466 million dollars) was divvied between research (227 million dollars), prevention programs (137 million dollars), and education/public awareness (60 million dollars). Habitat restoration efforts pulled the smallest support (42 million dollars).[9] This gap in funding exposes the invested interests and the inferior priority placed upon saving native species, the purported victims of these invasive plants, habitat conservation, protecting endangered plants, and restoration of wild places. Many of the plants that abundantly inhabited our woodlands, prairies, wetlands, and other unique environments now struggle for existence due to a variety of factors, and attempts to restore their way of life are grossly underfunded.

Unfortunately, endangered North American plants also receive inferior protection under the federal Endangered Species Act and other laws. According to the Native Plant Conservation Campaign:

> Nearly sixty percent of species listed under the Endangered Species
> Act are plants, but less than three percent of federal endangered species
> funding goes to plants. . . . One example is the federally funded
> Wildlife Action Plan program, which provides money for state species
> and habitat conservation projects. More than $400 million was disbursed by the program between 2001 and 2006, but not a dollar went
> to plants since federal law explicitly prohibits states from using Wildlife
> Action Plan funds for plant conservation.[10]

The report goes on to explain that the federal Endangered Species Act allows some of these endangered plants to be "knowingly driven to extinction without violating the Federal Act" by allowing some listed species to be killed, without limits, on nonfederal lands. This is in stark contrast to the protection of animals listed with the act. For them, the act prohibits "the unauthorized destruction or even harm [of them] everywhere they occur."

Endangered plants are:

> only protected (1) on Federal lands, or during activities that are funded, permitted, or carried out by a Federal agency and are therefore under Federal jurisdiction, or (2) in the unlikely event that it can be proved that they are destroyed in knowing violation of state law or during trespassing. . . . [Therefore] logging, housing development, mining, and other activities may all kill unlimited numbers of federally listed plants, even cause extinction of a species, as long as the destruction does not meet these conditions.[11]

In addition, dozens of passed and pending bills have gone before the U.S. Congress to address and regulate so-called invasive species that have irritated a different crowd: private property owners. Some legislation allows the government to begin "invading" private property in order to combat so-called invading plants. Under such circumstances, the government further regulates land and water use and influences farmers, ranchers, and other landowners to abide by its wishes. According to Dana Joel Gattuso, senior fellow at the National Center for Public Policy Research in Washington, D.C.,

> Most bills would expand federal authority to further control land use and authorize billions of tax dollars to eradicate species . . . not based on science but rather assume all non-indigenous species are harmful unless proven otherwise. Most of the bills would create massive government bureaucracies and, worse, grant federal agencies greater authority to regulate lands and waters, public or private, where these species exist . . . and would lure private landowners with monetary incentives to assist in the effort, setting the stage for increased government involvement in land use on private property.[12]

The attorney for the California Farm Bureau, Michele Dias, echoes the dangers to farmers and private property owners of federally regulating invasive plants. She says, "Unless farmers and ranchers become active in their approach to this issue now, due to heavy environmental influence, federal controls could far surpass the type of abuses of power already experienced with the Endangered Species Act."[13]

A variety of factors have led to the present-day invasion, with government bureaucracy setting the standard. The founding fathers of this country had different views for the public good than today's children of these fathers, all of whom have their own vested interests.

> *The former set out to create a country, and the latter, seemingly at times, set out to destroy it.*

Vast amounts of money are plundered by various government agencies who write the policies that create a good-versus-evil dichotomy. The War on Invasive Plants is merely one such dichotomy that has been created from this mentality, and the government has been quite successful in its efforts, limiting the stage to a one-actor diatribe. The scientific authority piecemeals the studies of invasive biology to suit the policymaker's objectives, and it does not allow for further discussion among ecologists who view environmental changes differently—and this has led to domination by the instigators.

> *Again, the definition of an invasive plant: "An alien species whose introduction does or is likely to cause economic or environmental harm or harm to human health."[14]*

To explore further the issues surrounding these plants, I will use the general definition set forth in Executive Order 13112, (1) because of its basis for policymaking (and thus access to vast amounts of money), and (2) because of the order's creation of a social belief system. The very definition of an invasive plant leads to an onslaught of questions with unconvincing answers by the so-called specialists. The following chapters set out to examine these points and to broaden our understanding of these plants and their effects on environments and humans alike:

- What is meant by *native?*
- What is the difference between an *alien species* invasion and *native species* that invade?
- Is there a difference between human introduction and other natural dispersal?
- What are the underlying causes of these plants' movements?
- How is *environmental harm* scientifically validated?
- What is the *economic harm* that these plants impose, and against whom?
- What is the potential economic and environmental value of these plants, and how does this compare to the costs of mass eradication efforts?
- Are these plants actually harmful, or could they be healing for humans and the environment?
- What are the future impacts of our present actions against these plants?

Would the New World's colonizing European settlers fit the definition of an invasive species?

Some of the great hysterics about invasive species:[15]

We are experiencing an invasive species crisis. Invasive species will take over America's wildlife refuges, unless we act now.
NATIONAL WILDLIFE REFUGE ASSOCIATION

One of the greatest threats to the Earth's biological diversity. America is under siege by invasive species.
INTERAGENCY TASK FORCE OF
FEDERAL GOVERNMENT SCIENTISTS, 2005

Ecological mayhem caused by nonnative species.
DAVID QUAMMEN, INVASION BIOLOGIST

An insidious and pervasive conservation problem.

DANIEL SIMBERLOFF, INVASION BIOLOGIST

These are like something from a bad horror movie.

FORMER U.S. SECRETARY OF INTERIOR GAIL NORTON

All [exotics] should be treated as threats . . . unless proven otherwise.

PATTEN AND ERICKSON

The single most formidable threat of natural disaster of the 21st century.

NASA

2
The Science of Invasions

The living and holistic biosystem that is nature cannot be dissected or resolved into its parts. Once broken down, it dies. Or rather, those who break off a piece of nature lay hold of something that is dead, and, unaware that what they are examining is no longer what they think it to be, claim to understand nature. . . . Because [man] starts off with misconceptions about nature and takes the wrong approach to understanding it, regardless of how rational his thinking, everything winds up all wrong.

MASANOBU FUKUOKA

WHEN THE TREMORS CREATED by plant outsiders were just beginning to amplify in the 1940s and 1950s, and the fear of foreign invasion from World War II was on everyone's mind, a whole new field of ecology called *invasion biology* sprang forth from the work of Charles Elton and his seminal piece, *The Ecology of Invasions by Animals and Plants*. Elton's research covered all invasive organisms—plants, animals, insects, microbes—presenting the ideology that invasions are considered harmful changes to ecosystems by the successful establishment and colonization of a new species outside its known range.

Seen from a different perspective, the development termed *invasion* could also be described as a "vegetation dynamic" or "successional change" that is a natural process of plant species and ecosystems to deal with disruptions and openings. The majority of plant invasion theorists agree that the first cause of invasion is associated with the opening of a niche within the environment: these plant pioneers settle on the disturbance. Beyond this, however, many scientists provide inadequate explanations about the following complex, intertwining processes that involve hundreds, if not thousands, of species. Many reports fail to remember the original disruption and, with limited understanding and isolated viewpoints of multifaceted ecosystems, go on to describe how the plants perform. This shortsighted version of ecological dynamics and plant relationships has led to many misconceptions and incomplete theories that cannot properly explain the situation.

The science of invasives appears to be "ideological rather than evidence based,"[1] as David Theodoropoulos states, where the very words used to describe these plants and the definitions on which scientists base their theories have little scientific standing and create much debate about their actual meaning. The very notion of using the term *invasive species* "lays the blame for invasions squarely at the feet of the [plants] themselves,"[2] as permaculturalist Dave Jacke has noted, and doesn't consider the greater picture. There is muted debate between the different fields of ecology and inadequate analysis of how invasions occur, how to predict them, and what to do about them. A wide breadth of information and convincing arguments made by researchers that discredit the common invasion science is available but rarely finds its way to the public and government discourse.

The basic science of invasions is where we can begin to understand our dilemma.

INVASION AND THE SUCCESSION OF PLANTS

We have to accept the proposition that invasions of animals and plants and their parasites—as well as our parasites—will continue as far as the next Millennium and probably for thousands of years beyond it. Every year will see some new development in this situation. That is a way of saying that the balance between species is going to keep changing in every country.

CHARLES ELTON, *THE ECOLOGY OF INVASIONS BY ANIMALS AND PLANTS*

Imagine a forested landscape that has recently been cleared of all the trees and plants that have commingled there for hundreds of years. Bare earth is revealed, watercourses changed, and all species, both visible and invisible, have felt this trauma. Delicately layered soils are fractured and opened. They then have to make adjustments to cope and begin to rejuvenate the environment that has been upturned. The clearing slowly begins to turn into fields of pioneering plants, eventually making way for shrubs and saplings, culminating with the return of forest, which takes hundreds of years to mature.

As the weed watcher knows, nature does not wish to expose the soil; it always tries to cover it with plants. These plants fill in all spaces that provide sunlight; they access different degrees of brightness on all planes of plant structure, from the soil to the forest canopy. Around the edges of a clearing, shade-tolerant saplings and vines quickly sprout, taking advantage of the circumstances. Each has an innate form and function to help fill in the gaps and rehabilitate the devastated area. Often, an invasion of plants occurs in such circumstances: certain resources that benefit them—such as light, nutrients, and water—are more plentiful, and most opportunistic plants have the capability to find and optimize these essential resources. These plants also propagate readily and freely, with highly adaptive capabilities, and they tend to thrive and expand into disturbed or altered ecosystems. The incredible ability of pioneering species to cover wounded landscapes in such prolific numbers makes them easily misunderstood as invasive intruders.

The succession of plant pioneers in the unfolding ecosystem prepares the soil for other species to follow, first by protecting land from further erosion; then by enriching the soil with large quantities of biomass and providing essential nutrients with uptake capabilities; and, finally, by balancing microbes in the soil ecology. These plant pioneers essentially create

life out of destruction—life with striving interdependence and coopera-
tion between all species within the ecosystem (plants, animals, insects,
and microorganisms).

Understanding plant community dynamics and species niches is fundamental to understanding invasive plant motivations within ecosystems.
First, it is fairly clear that in healthy and whole ecosystems, all plants
share a common purpose: to maintain the optimal conditions in which
all species can thrive and evolve in a self-renewing and self-sustainable
fashion. Each plant has a key role to play that supports the whole community, much like the jobs people fulfill to sustain a village. Each individual
provides an essential purpose to the continuity of the community's life,
and with this, a niche is filled and the collective becomes more whole. In
understanding the purposes of any plant, Dave Jacke notes that "the species niche defines the unique characteristics, behaviors, and adaptations
of a particular species,"[3] and the surroundings of each plant reveals this.
Each plant species is an individual serving a purpose for the land by gath-
ering energy from the sun and soil, transforming it, and then providing it
for further use by the other inhabitants of the community.

> *Ecological communities are not as tightly linked as organisms,*
> *but neither are they simply collections of individuals. Rather,*
> *the community is a unique form of biological system in which*
> *the individuality of the parts (i.e., species and individuals)*
> *acts paradoxically to bind the system together.*
>
> DAVID PERRY, *FOREST ECOSYSTEMS*

All of an individual plant's characteristic functions and abilities must
be taken in context with the whole ecosystem that it inhabits. An invasive
species must be understood from this standpoint in order to realize its
environmental niche and the availability of resources and openings due
to disturbance, development, and contamination that further helps such
an adaptive species to inhabit and spread throughout the land. Even with
wide consensus that disruptions to ecosystems are altogether too common, such disturbances have taken place, resulting in the fact that oppor-

tunistic species are endemic—though this circumstance is most often overlooked in the further battle against invasive plants and continues to lead researchers away from this very crucial and basic element to these plants' existence.

Often, the invasive plant landscapes—the seemingly healthy ecosystems that are colonized—are in fact under stress, and scientists and those in control fail to recognize that a disturbance is not always a visual wound. In addition to widespread physical alterations to landscapes, massive upheavals to ecosystems occur from contamination by numerous invisible pollutants that have leached into the water, soil, and air. In such disturbed ecosystems, many of the native plants are poisoned and are less able to deal with upheaval, but the weedy, invasive plants cope well and even flourish in the toxic surroundings. Any disruption affects the plant community in drastic ways, changing shade and sunlight, soil nutrients and moisture, runoff, and the basic integrity of all dependent life-forms. In some cases, such disruption can be likened to a bomb exploding in a village, damaging all infrastructure and leaving dangerous residue and rubble and other long-lasting negative effects.

After the barrage has passed, a specially equipped group of plants are the only ones capable of remaining, and these serve as a sort of hazmat team until all is safe for others to return. These plant species, in this way, are barometers for the health and status of their environment. In turn, we can learn much about these plants themselves by studying their habits. The capacity to survive in disrupted environments describes the basic characteristics and functions of these plants, and at times we can measure their survival mechanisms—for instance, by testing for heavy-metal tolerances or soil conditions—but other times, we cannot quantify their abilities and must leave them to observation or intuition, or these strategies or abilities must remain unknown. The plants develop strategies in order to endure, and they use these adaptive qualities when others seem to fail. We can see these adaptations in the form and structure of each plant, both above ground and in the roots. This signature of the plant tells much about what habitats and stages of succession it occupies, how the plant acquires nutrients and water, and how it affects and interacts with others species within the environment—essentially, sending its own *signature message* from where it lives.

> *An organism's strategy constitutes the core of its species niche—*
> *how it makes its living. It unifies and organizes the disparate*
> *details of the organism's tolerances, preferences, needs,*
> *and yields into a coherent whole. It reflects the organism's*
> *evolutionary "choices" about how to spend its energy to adapt*
> *to its environment.*
>
> DAVE JACKE, *EDIBLE FOREST GARDENS*

Ecological systems are constantly changing. They experience highly unpredictable fluctuations in flora populations, and a diversity of species fill a variety of niches. Plants move throughout expansive ranges, sometimes with dramatic speed to take advantage of great disruptions. The time scale of these invasions are unknown, and the degree of their establishment is never fixed. Projecting our competitive nature, we humans tend to believe the exotic invaders are out-competing the natives, but it is not that simple. The flourishing of pioneer plants on damaged lands leads to these plants' value as healers of the land, thereby creating a new template of life-forms that has never been seen before. Regarding them as *recombinant ecologies,* permaculture co-originator David Holgren credits these plants with essential functions to renew and restructure broken ecosystems.

> *It is those who believe only in science who call an insect [or*
> *plant] a pest or a predator and cry out that nature is a violent*
> *world of relativity and contradiction in which the strong feed*
> *on the weak. . . . These are only distinctions invented by man.*
> *Nature maintained a great harmony without such notions,*
> *and brought forth the grasses and trees without the "helping"*
> *hand of man.*
>
> MASANOBU FUKUOKA,
> THE NATURAL WAY OF FARMING

THE UNPREDICTABLE NATURE OF NATURE AND THE EVASION OF INVASION BIOLOGY

Though ecologists claim to understand invasive plant ecosystem dynamics, they admit it is impossible for them to predict which plants are likely to invade, where they might invade, and how they might impact the environment. These failures undermine the assertions that many ecologists make. Invasive species expert Daniel Simberloff writes, "[V]irtually every specialist in invasion biology who has examined the matter concludes that aspects of the ecological impact of a non-indigenous species are inherently unpredictable," and he adds, "the effects of introduced species are so poorly understood and the record of predicting which ones will cause problems is so bad that one can question how much credence to place in a risk assessment."[4] Examples abound, with reports of plant populations unexpectedly booming after years of being in check. Smooth cordgrass (*Spartina alterniflora*), a native of the U.S. East Coast, was present in small patches along the Pacific coast for at least fifty years before it rampantly colonized salt marshes there. Charles Elton presents a case study of invasive plants' unpredictable natures when he describes Canadian waterweed (*Elodea Canadensis*), introduced into Great Britain in the mid-1800s, and its explosion "into rivers, canals, ditches, lochs and ponds all over the country, . . . [so much so that] . . . fishermen could not operate their nets. . . . [A]t Cambridge it clogged the River Cam, interfered with rowing . . . and render[ed] parts of the Thames impassable."[5] Then after twenty to thirty years, "it declined considerably and universally, and has never again been considered a real plague. . . . The reasons for its decline are quite unknown . . . but one thing is quite certain: *man did not directly control this weed*" (my italics).[6] This admitted failure to understand the basis of invasive plant ecosystem dynamics leads to further misunderstandings throughout the science of invasive biology and presents as the norm the evil of these plants.

> *I hate the "exotics are evil" bit, because it's so unscientific.*
> DOV SAX, ECOLOGIST AT BROWN UNIVERSITY,
> IN "FRIENDLY INVADERS"[7]

This misinformation, coupled with subdued discourse among other researchers in the fields of ecology, means that invasion biology stands at the fringe of science and fails to weave together the understanding from peers with differing perspectives. M. A. Davis, K. Thompson, and J. P. Grime point this out: "There currently exists an enormously rich literature of succession ecology that is being virtually ignored by many researchers studying invasions."[8] Also disregarded are the fields of evolutionary biology, plant ecology, paleoecology, bioremediation, plant pharmacology, climate change, and soil ecology—despite their individual relevance to nearly every plant species distribution and invasive situation.

> *Natural scientists don't like to study invasives. They consider them a kind of ecological pollution.*
> CHRIS HANEY, CHIEF SCIENTIST FOR DEFENDERS OF
> WILDLIFE, IN "INVASIVE SPECIES"[9]

When Elton began his research into invasive species, he started with the human perspective of the difficulties that these species caused. As a scientist, in order to fix these "problems," he used the standard scientific analysis to understand them. This mechanized worldview that most scientific endeavors embody details the simple pieces of the situation without revealing any understanding of the multilayered, multidimensional aspects of an ecosystem. This isolation of circumstances by focusing on only these "bad" plants instead of the whole picture leads to the very data that verifies the original difficulties. Such narrowness of thought leads to a great amount of detailed information that is valid in its own right, but that is still just part of the story. There are many other perspectives that can be seen from different angles. If we step back to see all of the information, we can gain a truer understanding that incorporates the unique standpoints of dozens of scholars in the fields of the biosciences and the plant world.

An in-depth, far-out way to look at ecology . . .

Originating From:

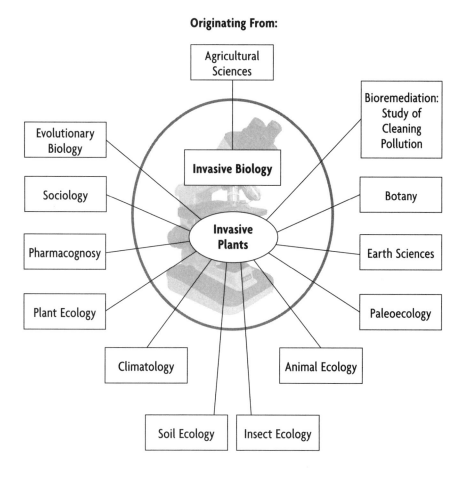

Scientifically, invasive plants are typically viewed through the narrow focus of invasion biology; however, if you step back and look at invasive plants from a broader perspective, it's clear that significant insight from many other fields of study can bring to light the true nature of these plants.

So lack of collaboration and prejudicial judgments permeate the science of invasions, leading to evaluations based on false premises. The following chapters are an attempt to bring together seemingly disjointed aspects of science in order to create a more comprehensive understanding of the ecological processes involving these misunderstood plants.

But first, some psychology . . .

EMOTIONALISM AND
IDEOLOGY

When scientists start saying that this plant is "bad" or "good"
they are imposing their opinions and not their observations.
When policy makers start making policies based on opinions,
we suffer as a nation.

ROBERT BEYFUSS, AGRICULTURE AND NATURAL
RESOURCES ISSUES LEADER, CORNELL UNIVERSITY

While he predominantly studied the pests that interfere with agriculture, Charles Elton, for the most part, kept standard scientific detachment throughout his work—although he still defended his position with tinted spectacles, being in close alliance with agriculturists. Unlike the fervent and hasty scientists of today, who wish to quarantine these plants, Elton did not wish to exclude foreign species on the whole. In fact, he writes, "I see no reason why the reconstitution of communities to make them rich and interesting and stable should not include a careful selection of exotic forms, especially as many of these are in any case going to arrive in due course and occupy some niche."[10] Although we have found it impossible to know the motivation of plants in various ecosystems, we continue a blockade and assault against these outsiders, showing deeply unscientific emotionalism and deeming the intruders guilty until proved innocent. Indeed, the derogatory language used to describe these plants has emotional undertones that have no place in scientific discussions. These deep-seated feelings can lead to misconstrued interpretations that are antithetical to the standard scientific analysis. They generally result in projections and isolation of data to prove a point that tells only part of the story. The very words *alien, noxious, invasive, aggressive, harmful, disruptive, choking, polluting,* and *villainous* cannot describe a scientific, ecological understanding of plant dynamics within complex ecosystems.

*Elton used a plethora of dramatic examples and vivid
metaphors to . . . encompass many disparate phenomena.*
DANIEL SIMBERLOFF, FOREWORD TO *THE ECOLOGY OF
INVASIONS BY ANIMALS AND PLANTS*

*I take this to mean that Elton wished to make a point using
all means possible. Well . . . I guess that is what I'm trying to
do, too.*

The basic ideology that creates the dichotomy of a native plant versus
an exotic one and a "good" plant versus a "bad" one is riddled with a deep
societal prejudice against outsiders (*xenophobia*), which has been exhibited in many cultures throughout history. This has led to an *authoritar-
ian nativism,* which projects an enemy from outside and creates hysteria
and anxiety among the population in order to rid the plant extremists
with vigor, ritual, and violence. These outspoken nativists also have an
arrogance that adds gusto to their argument. These same puritanical acts
against outsiders can be seen throughout history, and the great despair
they have caused mirrors the present-day plant inquisition. Today's nativism is similar to the mind-set in Nazi Germany, which, predating Elton's
dissertation, pioneered chemical warfare against an "inferior" race of
green weeds in its pursuit of purism. Though the kingdom classification
varies from plant to animal (insect and microbe), the motivating psychology remains constant. This differentiated ideology leaves no room for the
changing expressions of nature that have been evolving the genetics of
biosystems and Earth's inhabitants for eons. We must look at the interconnections among all of us—plant, animal (human), and microbial—
and remember that we are all natives that share this home we call Earth
and that we are inherently bound to one another with spiraling threads.

*We do, anyway, share 95 percent of the same DNA sequence as
bacteria. And besides, does science really have an understanding
of how nature ultimately works? Remember, it is the same
linear science that isolated and encouraged the growth of most
of these plants we are now trying to eradicate.*

Theodoropoulos states it more bluntly:

The emotionalism, the superstitious fear of exotic species, the lack of operational definitions of fundamental concepts, the repetition of subjective and anecdotal observations, the prevalence of misattributed causality and unjustified generalizations, and the ideology's apparent lack of familiarity with basic paleobiology, ecology, and evolutionary biology all indicated that invasion biology was perhaps a pseudoscience.[11]

3
Naturally Native:
Plants on the Move

Nature is one. There is no starting point or destination, only an unending flux, a continuous metamorphosis of all things.

MASANOBU FUKUOKA,
THE NATURAL WAY OF FARMING

There is a long lost cry for something we believe to be native,
yet we will fail to find this place unless we change our way of
    perceiving,
and consider this great living planet as a whole, our native
    home,
with biological richness as the life blood,
and impermanence as the heartbeat.
Always changing, nothing stays the same . . .

SEEDS OF CHANGE

There is a vital spirit in every seed to beget its like; there is a greater heat in the seed than any other part of the plant; and heat is the mother of action.

MAUD GRIEVE, *A MODERN HERBAL*

Plants have been on the move since the first seeds ripened some seven hundred million years ago, and since the first growth of these green wonders, environments have changed—from a chlorophyll carpet of long-lost solar soakers to giant plant dinosaurs, all feeding into the genes of the present species diversity we know today. Throughout time, they have had to contend with pressures that might have posed some difficulties: plants had to get up and move, they had to interact with other species, and they experienced mass extinctions, along with all else that might be difficult in a plant's world.

Changes, however, have come naturally and sometimes quite rapidly. Plants have been dispersed by wind, water, animals, and humans (both purposefully and by accident). They have traveled far and wide, floating across seas and up and down mountains. The farthest known seed dispersal without the aid of humans is fifteen thousand miles. Dust storms can carry seeds, spores, and insects from the Sahara Desert to Texas, and ocean currents can carry seeds and spores across thousands of miles of open water to inhabit new islands and continents. Darwin once commented that he witnessed seeds within the soil of a tree root stump that had drifted across the ocean.[1]

And must Darwin have wondered, "How will this evolve?"

Birds have migrated north and south, east and west, across continents and seas, carrying seed-filled excrement, and for eons all animals have traveled to favorable climes. Yet in the invasive plant rhetoric, the experts don't differentiate between human-introduced invasive plants, naturally dispersed plants, and bird-initiated plant movement.

Plants have adapted dramatically to changes in landscape. The climate has been a roller coaster ride of hot periods and cold periods, drought and flood. Whole continents have moved, creating landmasses and bridges that today are hidden beneath the sea. Humans have jumped the islands of Micronesia, Australia, New Zealand, the Bering Strait, and the English Channel, and plants have easily followed. Plain and simple, plants get around. If we think of plants and their relationship to their surroundings, our short life span of, say, eighty years does not compare to the life spans of plant ecosystems that are thousands of years old. Just step back and view it from the vast timeline of Earth's billions of years of existence.

> *It's interesting to sit with Gaia, the ancient, ever-changing, abundant Earth, and with her perspective realize our existence is but a blink of an eye . . .*

WHAT IS NATIVE?

When most people think of *native,* they think of a time before the colonialization of North America about five hundred years ago. The division between *native* and *exotic* is artificial, determined by a date, with those showing up afterward deemed *alien.* Some imagine a long-lost time that can never be re-created; they forget that nature is constantly changing. I wish I could have sat among the giant trees of the ancient forest, and I am even trying to restore a forest where I live, but that long-lost time and place are merely imagined. Plants have been on the move for ages, now appearing, now disappearing, though some people believe the landscape to be fixed and unchanging.

> *As human observers who measure change in days and years, it is difficult for us to grasp that the landscapes we know in our lifetimes are not only ephemeral, but also often radically different from those that preceded them.*
> TOM WESSELS, *READING THE FORESTED LANDSCAPE*

Take, for example, my surroundings here in New England. Its changes have played out in similar ways throughout North America. Pre-1500, old-growth forests of white chestnut, oak, hickory, maple, and white pine grew. In the 1800s, forests were clear-cut and pastureland was dominant. Today, mixed young forests sprout from the fields. (In the American West, however, forests don't "sprout" as they do in the east.) Thus, the concept of native plants and ecosystems is arbitrary, encompassing what we believe to be native and not necessarily representing what has been reality. In fact, we will never know what was the native reality—and perhaps it is a false construct.

Native species *as defined by the Invasive Species Council:
"[W]ith respect to a particular ecosystem, a species that, other
than as a result of an introduction, historically occurred or
currently occurs in that ecosystem."*

*Which conveniently lacks the possibility of what may
become. . . .*

I am reminded of the trips I took through the Midwest
and the miles and miles of corn I saw. There was cornfield
after cornfield, with some individual grasses and probably
invasives dispersed along the roadsides. Corn and then more
corn. A plant with origins in Central and South America,
corn was brought to this continent with Native Americans a
few thousand years ago. Today, there are eighty million acres
of nonnative corn growing in the United States, along with
sixty million acres of alien wheat and seventy million acres
of exotic soybeans—all of which were transported with the
help of humans. (And don't forget about the hundreds of
millions of acres of common lawn and golf courses planted
with exotic grasses wherever there's space!) Within this vast
acreage, very few, if any, native species share the land. Yet
an estimated one hundred million acres across America are
inhabited by the pandemic plants that are vehemently mowed
by machines or sprayed by planes.

On one of my trips, in the desert of homogenized corn, I was
fortunate enough to find the oasis of a state park. It seemed to
levitate on the horizon, and it transformed my familiarity with
this vast, flat land. In the park, waves of grass and wildflowers
rolled in the wind while I meandered among oak trees lining
the welcome crevasse in the land. I felt free, walking in the only
growing wildness for miles around. I imagined that in some
ways it must be how the savannas in Africa might feel. That
night I had lucid dreams.

NATURAL CLIMATE CHANGES

Given that the climate has been undergoing rapid changes with high variability during the twentieth and twenty-first centuries, we would expect population demographics and species ranges to also be highly unstable.

CONSTANCE I. MILLAR AND LINDA B. BRUBAKER,
"CLIMATE CHANGE AND PALEOECOLOGY"[2]

Although the human hand has done a large share of moving plants around, the backdrop of most of the plant movement (and human movement, for that matter) has been natural climatic changes. Plants continually follow favorable climates and environments, and the weather has fluctuated between warm and cold, dry and wet, for the past 2.5 million years. Even within the past 2 percent of Earth's life, California was marked by a jungle climate with tropical flora that went as far north as Alaska on the coast and to the Canadian border inland.[3] Scientists know the earth goes through cycles of warming and cooling on greater and lesser magnitudes, and we can see patterns in one hundred thousand year–cycles, ten thousand–year cycles, thousand-year cycles, hundred-year cycles, and even ten-year cycles. The oscillating thousand-year cycles of warming and cooling are called Bond cycles and have been documented for the past one hundred thirty thousand years. These seem to follow closely the trend of the sun's intensity via sunspots and other changes on the sun's surface— alternating between times of extensive solar activity that warm the planet and periods of a relative absence of solar activity, which brings on ice ages. A study by NASA has tracked the increasing strength of the sun and its relationship to the planet's rising temperature over the past century, and it was discovered that the sun has accounted for at least 25 percent of a 1.1-degree-Fahrenheit increase.[4] This lends credence—as if we need any more—to the importance of our great star and its influence over our lives (and why many cultures have revered it as a god).

Although, as I write this at the end of 2008, I am informed of the complete absence of any major sunspots in the past year. Brr, it's getting cold . . .

We are presently (over the past century) in a warmer period in relation to the past four hundred years, and that will affect our plant communities. When the first settlers came to America in the mid-1600s to the mid-1700s, the Little Ice Age was upon them (and sunspots were rarely observed during this time called the Maunder Minimum [1645–1717]), and the ecological situation was very different from that of today. In setting, then, a premodern contact date to mark the changes that have occurred in our plant environments, we have failed to take into account the natural variability in the plants' preferred ecological and climatic conditions. Because plants can't get up and move, they migrate by dying in one location and moving into another, with sporadic expressions throughout the landscape until they locate a place with a certain degree of stability. To study the changes in biota, scientists have compiled evidence by analyzing pollen accumulation in wetlands, bogs, and lakes; fossilized deposits in deserts; tree-ring growth in temperate forests; and coral formation in the ocean. It has been found that plants have been able to move hundreds of miles north and south and thousands of feet up and down mountainous regions. In discontinuous landscapes, plants responded with patchy movement, varying in population size and minor changes in locale. When areas covered in ice began to melt, vegetation moved in rapidly, invading the barren ground. In some cases, such as that of juniper (*Juniperus*), which has been considered an invasive species in the Great Basin rangelands, plants have been adapting to the changing climate and reclaiming ancestral land they lived on a couple millennia ago.

Various species of trees have followed similar trends in making mass migrations. One example involves the spruce (*Picea*) forests of eastern North American, and other species have followed similar trends. Forests of spruce once spread deep into the south when great glaciers covered the north eighteen thousand years ago. As the temperatures warmed and the ice melted, the forests retreated northward hundreds of miles, and in the present, spruce occur only in the far north.

The giant sequoia (*Sequoiadendron giganteum*) of the Sierra Nevada Mountains in California represent another example of mass movement of a plant species. Currently, these stately trees grow in small, patchy groves

Spruce

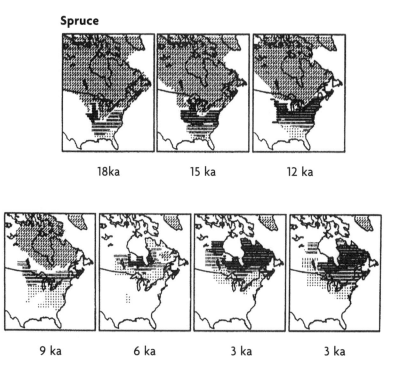

Shift in ranges of spruce (Picea) forests in eastern North America as they track changing temperatures from the Last Glacial Maximum to the present. Reconstructed from pollen abundances in lake sediments for intervals of three thousand years. From Constance I. Millar and Linda B. Brubaker, "Climate Change and Paleoecology: New Contexts for Restoration Ecology," in M. Palmer, D. Falk, and J. Zedler, eds., *Restoration Science* (Washington, D.C.: Island Press, 2006), 322.

in the southwestern range in a fairly isolated and seemingly restricted area (at an elevation between five thousand and sixty-nine hundred feet). Over the past ten thousand to twenty-six thousand years, however, the sequoia found hospitable living in the eastern range at Mono Lake and in the western range well above and below current elevation levels (up to 9,393 feet), down into current chaparral shrubland (3,300 feet), and into the California Central Valley at Tulare Lake (177 feet). The groves of today were established only forty-five hundred years ago, and the great giants are about two thousand years old, leading us to wonder where else this species might decide to get up and go.

One might say, this has happened over thousands of years, but such movement is happening so quickly now. Well, time does appear to have sped up. Trips across the world that used to take months or years now can be accomplished in a matter of hours, and people trot around the globe with their baggage filled with seeds of desire.

It is true that a span of thousands of years is difficult to comprehend in our decades-long lifetimes. During the past thousands of years, climates have varied immensely, and we are on the crest of a warming trend that started five hundred years ago. This warming has accelerated in the past half century, forcing plants and ecosystems to adjust in sometimes disturbing ways. Further, as plants adapt and move, animals, insects, and microbes are also adapting to these changing temperatures and thus affecting our environments.

Numerous studies show that climatic changes are at the helm of ecosystem alterations and apparent loss of plant species that have occurred in recent history. A study by the USDA confirmed this and found that the changing climate patterns, specifically increased snow loads in the prairies, helps facilitate current plant invasions. Opportunistic plants proliferate and increase biomass with the additional snow that these environments have experienced, while the native species have not responded so favorably.[5] Another study identified changes in tree species, forest structure, and forest biomass over the course of seventy-six years (1930–2006) in the Hudson Highlands of New York State. During this time of increasing temperatures, three species disappeared (American elm, paper birch, and black spruce) and seven species invaded the forest. Of the seven, two were nonnative (tree-of-heaven and white poplar), and the other five were native and were previously distributed farther south (southern catalpa, cockspur hawthorn, red mulberry, eastern cottonwood, and slippery elm). With this warming climate, the "understory tree community changed substantially . . . [and] there has been a significant loss of live tree biomass as a result of canopy tree mortality . . . [due mostly to] insect outbreaks and droughts."[6]

For those of us who farm, garden, or watch wild plants, we know every year is different: some plants are plentiful, and others seem to disappear; some years are high mast times with abundant fruit and nuts, and other years, pests swarm; some years are rainy and wet, and in other years, our gardens would shrivel if it were not for irrigation. Some plant species, such as ginseng, are known to hibernate underground for years if the conditions are not ideal, and seeds of most plants can lie dormant for years or decades until the environment meets the circumstances they need to sprout.

> *How many of the disappearing plants and their seeds do these things?*

Many of the movements and disappearances of plant species have been due to ongoing climate changes, and those policing the current situation have failed to incorporate this element of nature into the discussion of ecosystem transformation. We have also sometimes forgotten about the climate changes created by humans in the past century and how these relate to plant dynamics in the environment.

PLANTS ADAPTING TO HUMAN-INDUCED CLIMATE CHANGE

The weedy, opportunistic plants around us have strong abilities to adapt to the environmental and atmospheric changes brought about by humans. The effects of excessive carbon dioxide emissions have been said to lead to global warming, and if carbon dioxide emissions continue to climb, it's predicted that the earth will experience increased temperatures. The weed ecologist Lewis Ziska was interviewed by Tom Christopher in a *New York Times* article.[7] Working for the U.S. Department of Agriculture, Ziska set out to understand how the increase in temperatures and carbon dioxide might affect plant growth. He began searching for a laboratory to conduct his research and found it in nearby Baltimore, Maryland, which turns out to be a "heat island"—essentially, a concrete slab that absorbs the sun's energy.

> *Extensive concrete in the cities stores the heat and radiates the sun's energy back out into the atmosphere, thereby warming the air. This cement wrapping suppresses any life force from growing, effectively "plugging" every pore of the earth's skin, and does not allow plants to engage in the oxygen–carbon dioxide dynamic. Pavement inhibits any living soil microbial*

matrix and changes watercourses that channel and concentrate polluted runoff into nearby water sources. When we don't have plants in vast forests of skyscrapers to clean dirty emissions, the air is toxic for humans to breathe.

Temperatures in this city laboratory ran three to four degrees Fahrenheit warmer than the surrounding countryside, and due to pollution and lack of plant life to remove it by cleaning the air, Baltimore also matches the CO_2 concentrations that scientists predict for all of us thirty or fifty years into the future. This shows us the potential of what could happen to the planet as a whole by the year 2040.

Ziska set up three sites—an organic farm in the countryside, a park in a suburb, and a park in the industrialized inner harbor—to compare the impacts of elevated temperatures and CO_2 concentrations on some invasive weeds. The organic farm represented the present, the urban site represented the future, and the suburban site was representative of "somewhere in between." Ziska started with soil from the farm, which contained thirty-five common weed seeds, and created identical beds at each site. He then watched what happened over the next five years and was surprised by the results. "Not only did the weeds grow much larger in hotter, CO_2-enriched plots . . . but the urban, futuristic weeds also produced more pollen."[8] Lambs-quarters (*Chenopodium album*) on the farm grew an impressive six to eight feet in height, but in the city they were an alarming ten to twelve feet tall! Ziska also found that instead of a gradual return to woodlands taking place over many years, the succession of plants within the city plot was greatly accelerated, creating complete stands of invasive trees by the end of the five years. The largest tree-of-heaven on the farm plot was about five feet tall, but the urban-site ecology appeared to be "on amphetamines," unleashing a twenty-foot tree![9]

We are seeing one of the great historical convulsions in the world's fauna and flora.

CHARLES ELTON,
THE ECOLOGY OF INVASIONS

In another USDA study Ziska conducted, the increases in varying carbon dioxide levels in invasive plants were examined, and the results were similar to the field study—though the average increase in biomass growth was markedly greater over the past one hundred years compared to that projected for the next hundred years. Canada thistle (*Cirsium arvense*), bindweed (*Convolvulus arvensis*), leafy spurge (*Euphorbia esula*), sowthistle (*Sonchus arvensis*), knapweed (*Centaurea maculosa*), and yellow star thistle (*Centaurea solstitialis*) displayed a significant growth response (110 percent more growth) to increasing CO_2 levels during the twentieth century, with the growth increase anticipated to be an additional 46 percent over the next one hundred years. Overall, the weeds displayed three times more growth than any of the hundreds of crop and tree species previously examined for their response to increased carbon dioxide concentrations expected for the future.[10]

AGRICULTURE LENDS A HELPING HAND

From the time of the early radishes
To the time of the standing corn
Sleepy Henry Hackerman hoes.

There are laws in the village against weeds.
The law says a weed is wrong and shall be killed.
The weeds say life is a white and lovely thing
And the weeds come on and on in irrepressible regiments.
Sleepy Henry Hackerman hoes; and the village law uttering a
ban on weeds is unchangeable law.

CARL SANDBURG, "WEEDS"

From our need to understand our environment and how it is that our basic needs are met, it is easy for us humans to believe that nature is static, with relatively short life spans. This is why we have carried seeds and plants since our first migrations across the globe tens of thousands of years ago. The fifty-two-hundred-year-old Ice Man found in the Alps carried with him a medicine pouch with herbs and stuffed his shoes with grasses—none of which were native to the area where he was found. The movements of plants accelerated with the advent of agriculture and the promise for us of easier living, with decreased necessity for us to scavenge for food. For tens of thousands of years, humans have encouraged the spread of plants by carrying them for food, fiber, medicine, and beauty.

> *We are Nature, working. . . .*
>
> PENNY LIVINGSTON-STARK, PERMACULTURE
> DESIGNER, IN STARHAWK'S *EARTH PATH*

Even in hunter-gatherer societies, humans encourage the movement of plants intentionally—by planting ripe seeds after harvest—and by foraging portions of roots and shoots, which promotes regrowth. Then they encourage plant spread unintentionally: eating and pooping, moving and shaking all around.

During the advent of agricultural societies some eleven thousand years ago, human's undeniable instinct was to carry the potential of more food through carrying seeds. First crops gathered from wild wheat, barley, oats, and rice are the cousins of our modern invaders, and they share some of these invaders' characteristics. Whether through trade or natural migrations of people, these useful and edible grains have been transported in cargo to feed people and animals and have helped the plants expand hospitable territory. Due to natural variances within the grasses, a wheat grass was one of the first to be grown intentionally in the Fertile Crescent (Southwest Asia) around 8500 BCE. With its large seed head, it heralded a new interdependence between humans and plants.

Over time, a relationship has developed between people and the thousands of different species of plants used for food alone. Each of these plants has made a fantastic journey in both form and place, and many have taken over previously uninhabited land. Different cultures developed agriculture at different times, using a variety of wild plants. The approach of the first plant geneticists was simple (or perhaps lucky): a plant said,

"Look at my big seeds, look at how much there is of me," and the original plant scientists intentionally planted the seed of this plant and subsequently reaped a reward. Could, then, this situation be analyzed? Could early hypotheses be tested? For those less inclined to favor a native intelligence, the seeds were brought home and planted accidentally over and over—after which, they finally realized their potential.

> As a general rule, the rampant, loud plants with large seed heads offer much in return.

Or perhaps the most primitive people were the original agriculturists. Perhaps their methods went beyond our common ideas of agriculture—perhaps they closely resembled the permaculture movement of today: Original agriculturists created wild, edible gardens that reaped abundant yields on multiple levels of canopy. They thereby provided permanent harvests, and nature did most of the work. As Toby Hemenway points out, "Anthropologists mistook the lush and productive home gardens that enfolded tropical houses for wild jungle, so perfectly had the inhabitants mimicked the surrounding forest."[11] In order to create wild, edible forest gardens of trees and understory, native peoples throughout North America used permaculture methods—implementation of fire, selective harvesting, and intentional planting.

Back in the agricultural uprising, the Fertile Crescent valley in Southwest Asia was turned into an ocean of green wheat and barley that could feed the first civilizations, while China began cutting down the forests, about ninety-five hundred years ago, to make way for rice and millet paddies. In the Americas, corn, beans, squash, sunflower, tobacco, and others plants began to travel and establish themselves with the help of the people. Within generations, domestication of these plants spread throughout their respective regions, drastically altering wild states by displacing some plants with fields of other plants. Wheat, barley, flax, and so forth began their own migrations over land and sea, making journeys unknown to their ancestors—yet they adapted to all sorts of environments, and people helped these plants along the way by opening the land.

> Does this still ring true today?

ORIGINS OF THE DOMESTICATION OF PLANTS*

Wheat, barley, peas, olives	Southwest Asia	8500 BCE
Rice, millet, cannabis	China	by 7500 BCE
Corn, beans, squash	Mesoamerica	by 3500 BCE
Potato, manioc	Andes and Amazonia	by 3500 BCE
Sunflower, goosefoot	Eastern United States	2500 BCE
Sorghum, African rice	Sahel	by 5000 BCE
African yams, palm oil	Tropical West Africa	by 3000 BCE
Coffee, teff	Ethiopia	?
Sugar cane, banana	New Guinea	7000 BCE?
Poppy, oat	Western Europe	6000–3500 BCE
Sesame, eggplant	Indus Valley	7000 BCE
Sycamore fig, chufa	Egypt	6000 BCE

*Adapted from Jared Diamond, *Guns, Germs, and Steel* (New York: W. W. Norton & Co., 1999), 100.

Once we perfected the art of growing our own food from seeds and civilization began to flourish, we had more time for other gardening endeavors, so we brought closer to home the wild plants we praised for medicine and beauty. When we had to move our homes, we took these allies with us in order to have the reassurance that comes from knowing the plants around us.

> Today, some of us have lost our familiarity and comfort as new, exotic plants creep into our surroundings.

Of the plant crops we rely on for food these days in the United States, 98 percent are alien plants, with many developing from an invasive relative or becoming like an invasive monoculture crop.

Since the first upheaval of Earth's soil by the first farmers, uninvited seeds of other species have found their way into the garden. Thus, the emergence of the *weed* and our initiation into the prejudice against and preference for different plants. A weed exists only because of the discord between people and plants—and because we underestimate the power of coexistence among different plant species (and microorganisms and fungi) in the environment. The practice of weeding has contributed to the loss of millions of tons of topsoil to rain runoff and a reduction in soil fertility that has led to a cycle gripped by the use of fertilizer. The deeply penetrating roots of grasses and weeds help loosen the soil and bring nutrients to the surface of the earth as plants die and decompose. These dead plants add compost to the soil, feeding the next organisms in the cycle and enriching the earth—and thereby benefiting other plants in the vicinity. Masanobu Fukuoka pioneers the no-weeding, or "weed utility," principle in *The Natural Way of Farming,* which addresses other ways of thinking and acting like a farmer.

> In nature, plants live and thrive together. But man sees things differently. He sees coexistence as competition; he thinks of one plant as hindering the growth of another and believes that to raise a crop, he must remove other grasses and herbs. [Instead], monoculture is more convenient for the farmer [yet] it would be wiser to remove weeds with weeds than to pull weeds by hand by planting grasses or green manure crops that take the place of undesirable weeds and are beneficial to him and his crops, then he would no longer have to weed.[12]

> *I have done this myself with the overtrodden blackberry that fills the hillside down from our house. As I've worked my way down the hill, pulling and scraping my way along, I've placed on the ground thick layers of straw and mulch. On top, I've placed soil seeded mainly with white clover and some other legumes, chickory, and grasses that will inhibit most of the blackberry growth. I have found white clover to be a powerful land remediator. It energetically and physically melds the land with its strong root system, creating a matrix of nitrogen-fixing messages, integrating the "nerves" of the land, and renewing fertility to the soil. This eventually allows a species-rich, grassy environment to replace the brambled, overrun areas.*

HUMAN DISTURBANCES
BEHIND THEIR MOVEMENT

The twentieth century has eliminated the terror of the unknown darkness of nature by devastating nature herself.
THOMAS BERRY, *THE DREAM OF THE EARTH*

In addition to natural dispersal methods, climatic changes, and our ingrained impulse to carry the seeds of food and medicine with us, human-caused disruptions to environments have probably done the most in the past few centuries to pave the way for invasive plants to move to different areas. Every case of widespread invasion can be traced to the action of humans—and often, multiple factors contribute. Not only do human practices help spread these species, but also we are often the cause of habitat loss in the first place. Because invasives are more adaptable and prolific, they are quicker to seize the opportunity of open ground. Yet we continue to blame these innocent bystanders that simply fill the holes we humans are digging.

Human Disturbances Leading to
the Spread of Invasive Plants

- Roads, highways, trails, railways, and the electric grid
- Mining—including mountaintop mining—that brings radioactive or concentrated metals to the surface of the earth
- Logging
- Overgrazing livestock
- Intentional plantings
- Dams, canals, and water use
- Agricultural practices and soil depletion
- Pollution, including acid rain, herbicides, fertilizers, pharmaceuticals, landfills, and air
- Fire suppression and loss of Native American land management and tending

Roads, Highways, Trails, Railways, and the Electric Grid

The most ancient means in which weed plants use humans for dispersal is by way of roads and trails that have turned into the ever-widening highways of today. These plant opportunists are not picky about their method of moving: they hitch a ride on shoes, wheels, propellers, or dump trucks. Within the continental United States, fifty miles or less separates the wildest of wild places from a road, and trails for hikers, horses, and recreational vehicles spread throughout seemingly protected lands.

There are over forty thousand miles of limited-access highways in the United States. To build just one mile of interstate, occupying two 24-foot-wide lanes and a 50-foot median, forty acres of land is required at a cost of at least a million dollars. Each road has been built by excavating the earth with heavy machinery, which kills or displaces the original vegetation and opens a nook for foreigners. Oftentimes in the past, officials then intentionally plant these gloriously spreading, remediating invasives in order to stabilize the soil. In addition, the electricity grid established in the past century has led to large-track fissures, openings, and edges. Electricity now expands into the most remote areas, and this current carries seeds and plants during installation, repair, and maintenance jobs.

Mining

Mining practices have contributed not only to the spread of invasives but also to the utter destruction of environments, leaving them inhospitable to just about everyone. Whole mountaintops have been removed and placed in the valleys below.

Why has this been an acceptable practice?

Deep reserves of potent metals and radioactive compounds are extracted and processed into concentrations that the earth has never before experienced. Following this are cleaning and refining processes, which contaminate the mining area with toxic runoff. Very few species—especially the original species—can survive in such areas, but some plants such as phragmites and Japanese knotweed can thrive and can even help cleanse mining pollution. The scope of such operations is huge—which creates plenty of space and plenty of opportunity for such plants.

Logging

Since the time of the first settlers, logging practices have been much like a tsunami that has torn across the continent, leaving graveyards of giant stumps. Logging practices have created traumatized ecosystems that are most often turned into agricultural land and that have left most native species struggling for survival. Today, the mass machinery that is involved in this National Forest industry fractures the terrain, washes soil into the streams, and creates openings that must be filled. Trees are valued as wood, and the cheapest means possible to extract the commodity of lumber are implemented. Plants such as barberry and blackberry are quick to jump into the spaces created by improper forestry practices such as overthinning and taking the largest trees that offer greater canopy coverage. We must remember that by removing the largest trees, we weaken the overall structural diversity of a place. We must protect not only the biodiversity of species but also the biodiversity of longevity and the layers each species provides.

> *The elders in any community are the essential ancestral links who offer the stories to sprout, grow, and teach subsequent generations.*

Overgrazing Livestock

Livestock can tolerate a variety of circumstances, surviving desert extremes, the difficulties of high mountain land, and many other environments where there is some green roughage (and even if there's not). Yet they do considerable damage to fragile ecosystems, especially in the western states, where they trod on land that has little, if any topsoil. They also eat only certain plants and leave others to thrive. As early as the 1940s, William Vogt pointed out that mass degradation had already taken place on the American plains due to livestock. He found that by 1935 the capacity of the grazing herds on western rangelands had been reduced by half due to the serious damage to the vegetation by overgrazing and because most of the land was eroding, growing into an exspansive desert.

In addition, feedlots these days are a collective soup of excrement, hormones, and antibiotics that eventually overflows and makes its way into the groundwater and waterways.

Intentional Plantings

As mentioned earlier, many species of invasives have been intentionally planted in the wake of human-induced trauma in the ecosystem—and sometimes they have been planted en masse. A prime example is kudzu throughout the South, which the U.S. Department of Agriculture used extensively for erosion control. In fact, in the 1940s the department paid out eight dollars an acre to those who planted kudzu in old fields for forage. Since the 1930s, the U.S. government has promoted multiflora rose for erosion control, highway divides, living fences, and wildlife benefits. Russian olive was employed as a means of fast-growing windbreaks, for erosion control, for wildlife forage and shelter, and as a potential crop supplement. Tamarisk was once revered for its striking appearance, and it was planted throughout drought-stricken landscapes for ornament, for use as windbreaks, and for streambank erosion control. The list goes on and on: many other opportunists were brought to this country for medicine, food, decoration, fiber, and soil-enhancing capabilities.

Dams, Canals, and Water Use

The "damnation" of our rivers (as Edward Abbey would say) to feed water-diversion programs has led to great ecological changes. Canals have been built for hundreds of years, which easily spread seed and root fragments through these expansive pathways and into wetlands. Massive lakes now occupy once towering canyons in order to provide drinking water for people, irrigation for crops, and water entertainment and recreation. This is especially notable in the American West, where water tables have dropped to dangerous levels for desert inhabitants. Snow-melt streams are eventually diverted to flush the toilets, water the lawns, and create pond-lined golf courses for hundreds of millions of people in the scorching desert sun. The plants have had to cope with these drastic changes in water level, leaving only the likes of tamarisk—those plants that have a greater capacity to reach a deep water table or to survive on less.

Don't forget about the motorboats that carry fragments of
water invasives from one damned lake to another.

Agricultural Practices and Soil Depletion

The settlers of this country tamed the wild land, beginning by plow-
ing the first fields. Precious topsoils began to blow away, seeping into
the waterways, forever leaving the Great Plains and washing up mostly
in the Gulf of Mexico. Then, during the Great Dust Bowl of the 1930s,
it drifted into the Atlantic Ocean. This produced depleted, infertile soil
that was perhaps suitable only for livestock forage, or, in some cases, was
left as hardpan or clay that was taken over only by certain types of plants,
such as the robust wild mustard, knapweed, and thistle.

> *Should many distinct groups of "wild" plants begin an invasion*
> *into an area where they have previously been absent . . . this is a*
> *sure a sign of decline in soils. . . . Were we attentive observers we*
> *would see that, above all, [weeds] are witnesses to our failure to*
> *treat soil properly and that their abundant growth takes place*
> *only where man has missed the point in regard to them. As*
> *nature's corrector of man's errors, they tell a silent story full of*
> *subtleties concerning the finer forces through which nature helps*
> *soils by balancing and healing.*
>
> Tompkins and Bird, *Secrets of the Soil*

As the farmers came to the new land and the plants came with them,
so did the earthworm, for it was praised in the Old World for enriching
garden soils. These creeping worms that we cherish in the rich black gold
of compost are an invasive species—as much as we love them—and could
be a huge instigator for the movement of invasive plants. According to an
article in *Conservation Biology:*

> These patterns suggest earthworm invasion, rather than non-native plant
> invasion, is the driving force behind changes in forest plant communi-
> ties in northeastern North America, including declines in native plant

species, and earthworm invasions appear to facilitate plant invasions in these forests. Thus, a focus on management of invasive plant species may be insufficient to protect northeastern forest understory species.[13]

This is one more example into underlying conditions that lead to the spread of plants, the decline of the native plant diversity, and the prejudice toward and preference for different species.

Pollution

The great pollution epidemic of our time and its perils could fill volumes of books. Herbicides, pesticides, chemical fertilizers, petroleum, pharmaceuticals, acid rain, and so forth: all are increasingly abundant, and no place is left on earth that doesn't feel their impact. Here I focus on the silent poison showering down on us, the plants, and ecosystems as a whole. Acid rain (mainly caused by coal power plants, which supply about half the nation's electricity and most of China's power) releases nitrogen compounds that collect in the soil at three times the normal amount. This accumulation of nitrogen in the soil leads to rapid plant growth and a reduction of lignin content in woody plants. Lignin accounts for the strength of woody plant tissues—and its absence leaves the plant more susceptible to disease and insects.

> *And I might add ice as a hazard for the trees who are burdened with acid rain and loss of strength. As I write this, trees all around me are snapping from the couple of inches of ice they collected overnight. What's more, my well water tested acidic because of pollution—and according to the water tester, nearly everyone in New England has acidic water in the well.*

This high-nitrogen environment also enhances the growth and vigor of opportunistic plants, helping them to overcrowd areas. The process by which our deeply toxic environments soon support the spread of invasive plants is actually twofold. The first step involves the manner in which this toxic surrounding impacts the native species that exist in broken environments and fragile ecosystems. Next, after the death of natives, these weedy cousins move in. They are more adaptable, and in some cases they actually thrive in excessively nitrogen-rich environments. We will focus more on pollution and its impact on plant ecosystems in chapter 7.

Fire Suppression

In many ways, Native Americans were stewards of their land, setting periodic fires to enliven the soil, retard pests, and stimulate the growth of certain plants. The departure of these people and their practices, along with the colonizing mind-set that incorporated avoidance and ignorance of the subject of fires that are purposely set and the continued suppression of wildfires, has led to a dramatic altering of ecosystems. The old, native places are being out-adapted by new ones. In fact, within fifty years of the mass settlement of New England, most fire-managed prairie lands along the coast (some forest edges were said to occur six miles from the sea) began to recede into woodland. With this came the loss of a uniquely managed ecosystem and the extinction of individual species such as the heath hen, a relative of the grouse, and potentially other plants and animals. Fire was a tool used by native tribes all across the country. It created and maintained vast prairie lands and grass savannas that teemed with wildlife; it opened the forest floor by removing bushes, brambles, and small trees; and it defeated mosquito breeding grounds, therefore allowing wildlife and people to reach an abundance of grasses and herbs and allowing clearer terrain for hunting. Certain fruits such as blueberries were revived with fire, and the dominant nut-bearing trees were fire resistant and provided plentiful food for all. Ecologist Mark Wilson writes, "As climate turned cooler and moister 4,000 years ago, oak savanna and prairie ecosystems [of the Northwest] were maintained only by frequent fires set by native people to stimulate food plants and help in hunting."[14]

> *I have witnessed environments drastically altered by other means before my very eyes. When my wife and I moved to our land, we enjoyed walking through and around the bogs that surround us, and we found intriguing plants that could be found in no other environments. Even some rare and endangered plants such as sundew, orchids, and Venus flytrap seemed to flourish in these belching carpets of green cushion (although Venus flytrap is*

reported to grow only farther south). We were captivated by these treasures, and we felt honored to be their neighbors. Then an invasion took place and we watched the disappearance of nearly all of these special plants. In one bog, then another, plants and trees vanished. We could do nothing—it happened so quickly.

It was the beaver that moved in, and now the bogs are ponds; the trees are dams, dens, and food; and the plants are under water. We grieved the loss, but it quickly turned in to joy, wonder, and honor—we witnessed an amazing transformation. In one season, we experienced nature's bulldozer, and we welcomed our new neighbors. This lesson of nature's impermanence happened directly outside of our door, and it is just one transformation of the environment in the many expressions of an ever-changing world.

As we have seen, *native* is a state of mind and a relative concept. Plants have moved in drastic forms and by various means—and there are no boarders in the plant world. Humans, the most complacent species, assists in this process, happily turning up the soil and spreading seeds. The most persistent plants continue to follow disturbance—and then we blame them for harming the ecosystem's inhabitants and upsetting the native environment.

Once we were gone, the prairie should have settled back into something like its natural populations in their natural balances, except. Except that we had plowed up two hundred acres of buffalo grass, and had imported Russian thistle—tumbleweed—with our seed wheat. For a season or two, some wheat would volunteer in the fallow fields. Then the tumbleweed would take over, and begin to roll. We homesteaded a semi-arid steppe and left it nearly a desert.

Not deliberately. We simply didn't know what we were doing. People in new environments seldom do. Their only compulsion is to impose themselves and their needs, their old habits and old crops, upon the new earth. They don't look to see what the new earth is doing naturally; they don't listen to its voice.

WALLACE STEGNER, *AMERICAN PLACES*, 1981

4

The Nature of Harm, the Harm to Nature

Often invasives are defined as harmful simply because someone has decided that any change caused by a nonnative species damages the native ecosystem. If harm is defined as change, then the question is why change caused by an alien is harmful and change caused by a native is natural?

S. R. KAUFMAN AND W. KAUFMAN,
INVASIVE PLANTS GUIDE

THERE HAS BEEN MUCH contention over the assertion that invasive plants harm the environment. They are said to push out natives, threaten wild lands, alter habitats, and reduce biodiversity. The Invasive Species Council defines environmental harm to mean "biologically significant decreases in native species populations, alterations to plant and animal communities or to ecological processes that native species and other desirable plants and animals and humans depend on for survival."[1] The National Park Service calls nonnatives "one of the greatest threats to our natural and cultural heritage,"[2] and the Union of Concerned Scientists considers the intruders "one of the most serious and least-recognized tragedies of our time."[3] Nonnative plant invaders are said to "infest" an estimated one hundred million acres, aggressively displacing native species and being one of the greatest threats to our environment. NASA condemns them as "the single most formidable threat of natural disaster of the 21st century,"[4] while at the same time they have been successfully researching the use of water hyacinth as a wastewater treatment plant, reporting that this invasive species that overtakes waterways in Florida "flourished on the sewage and the once-noxious test area became a clean aquatic flower garden."[5] This same process of using certain invasive plants to clean contaminants from water supplies is being implemented by numerous sewage treatment plants across the country and around the world—yet invasives are still called the plants that do harm.

First, this question must be asked: How do we scientifically validate and properly define *environmental harm*? Some contend *harm* to be any significant change to an ecosystem brought on by an alien species, yet as we have seen, a significant change can occur for a variety of reasons, both visible and invisible, human and nonhuman. Many native species—such as goldenrod, aster, and American blackberry—are flourific colonizers of disturbed land, yet this large-scale movement of native plants is seen as a natural succession of the ecosystem. North American plants such as Jerusalem artichoke (*Helianthus tuberosus*), groundnut (*Apios Americana*), miner's lettuce (*Montia perfoliata*), and fireweed (*Epilobium angustifolium*) are considered vigorously expansive and highly prolific, but when a foreign plant enters the equation, its spread is seen as regression and a step backward for the environment. Then there are the native poisonous plants, with their allelopathic compounds that keep at bay other plants such as North American black walnut (*Juglans nigra*), Maximilian sunflower (*Helianthus maximilianii*), and hackberry (*Celtis* spp.), which exude chemicals that kill surrounding vegetation. Yet many consider this simply what they do—these are antisocial species, but unlike the alien intruders so maligned for their toxic tendencies, they are not evil, bad plants that are out-competing the others. Let us remember the human projections that blind our way of thinking and leave us susceptible to a limited, dualistic mind-set.

BIODIVERSITY DEBATE:
CREATING EXTINCTIONS

It is unlikely that empirical studies, if undertaken, could show that in randomly selected ecosystems, non-native species, especially plants, are more important factors in extinction than are native species and many other contributing causes and conditions.

MARK SAGOFF, PH.D., SENIOR RESEARCH SCHOLAR AT
THE INSTITUTE FOR PHILOSOPHY AND PUBLIC POLICY AT
THE UNIVERSITY OF MARYLAND

What is widely known is that so-called invasive plants are opportunistic fellows who flourish in disturbed lands. According to the United States National Arboretum (USNA), "[I]nvasive plant species *thrive where the continuity of a natural ecosystem is breached* and are *abundant on disturbed sites like construction areas and road cuts.* Even foot traffic can create a temporary void that is quickly invaded—some national parks have restricted the areas where visitors are allowed to walk" (my italics).[6]

> So what have we done since the mid-1900s is cut endless roads, improperly manage lumbering, and mine the mountaintops throughout our national forests. We have allowed strip malls built on fragile wetlands and have expanded our highways wider and wider. We humans are the ones who are actually disrupting natural habitats.
>
> Has there been much caution with regard to our perpetual involvement in the creation of our perceived problem? How many logging roads have been built since the Invasive Act? How many disturbed wetlands? How many widening miles of highway, and how much extending of trails for off-road vehicles?

In the same discussion of invasive plants, the USNA came to a conclusion as to why they are a problem: "[I]nvasive plants alter habitats and reduce biodiversity."[7]

Now which is it? Are the plants taking advantage of the human-caused breach and destruction or are people themselves altering habitats and reducing biodiversity?

Ecologists continually state with alarm that invasive species are the second greatest threat to biodiversity worldwide. What they fail to mention or sometimes gloss over is that this threat comes *after* the threat of habitat destruction. This is a convenient change of focus from the main source of the problem.

How much energy and and how much of our government resources are spent on the "War on Habitat Destruction"?

Governments, backed by many people and corporations, have long been compliant and have reaped the huge benefits of habitat destruction (a.k.a. *development* and *industrialization*), and therefore have threatened ecosystems more than any plant. After five hundred years of colonization by nonnative Americans, less than 5 percent of the native, old-growth forests remain in the United States, leading to devastated and traumatized ecosystems on nearly 95 percent of the land, with almost every native species that has commingled in unique environs for thousands of years feeling the blow. People have clear-cut forests, dammed rivers, mined the earth, and acidified the rain, polluting every resource in the environment. Our soil is altered, our water is toxic, our air is smogged, and plants in fragile ecosystems are the direct recipients of these traumas.

The Environmental Protection Agency (EPA) outlines the "Threats to Wetlands," describing the dire circumstances we are in, given the destruction of more than half the wetlands that were thought to have existed in the lower forty-eight states since the 1600s. Of the original two hundred million acres of wetlands, aproximately one hundred million acres remain—and there was an alarmingly accelerated pace of loss occurring from the 1950s to the 1970s. Even today, every year, sixty thousand acres are still lost. Wetlands collect and purify water, provide floodwater storage, and reduce soil erosion from runoff. They are vital habitat for numerous plants, animals, creepers, and crawlers, and the enormous loss of these unique ecosystems has affected countless life-forms. All of this human-generated destruction began with draining wetlands for farmland, thereby revealing rich, fertile soil. Then the cookie-cutter developers moved in—with no oversight—and they believed, and continue to believe, that buildings are more important than bogs. Wetlands have also been damaged and drained for mosquito control, canal dredging, lake formation, livestock grazing, and peat-moss mining. Of the wetlands that remain, most are overburdened, coping with more toxicity and polluted inputs while garnering less support for purification from destroyed networks of wetlands. Generally the lowest-lying places within the landscape, wetlands are the collection tanks for springs, streams, and water runoff. They store and purify water before it makes its way into rivers and larger bodies of water. According to the EPA,

> The primary pollutants causing wetland degradation are sediment, fertilizer, human sewage, animal waste, road salts, pesticides, heavy metals, and selenium . . . originating from many sources, including: runoff from urban, agricultural, silvicultural [forestry], and mining areas; air pollution from cars, factories, and power plants; old landfills and dumps that leak toxic substances; marinas, where boats increase turbidity and release pollutants.[8]

The significant increases of pollutants has affected the diversity of flora within wetland environments, which has meant that what remains are plants that are adapted to increased pollution loads and those capable of readily migrating to the next available wetland. Opportunistic plants have offset the dwindling numbers of native plants whose original habitat has been encumbered with toxic accumulation from human actions and other sources. Japanese knotweed, for example, flourishes in abandoned copper mines and along polluted riparian areas, and, as it is one of the few plants or trees that can grow in such places, it enlivens and cleans the environment. Purple loosestrife, phragmites, and cattails also have heightened capacity to clean spoiled waters.

In native woodlands, plants such as ginseng, goldenseal, and other endangered species once blanketed forest floors and created a much greater diversity than is present in most forests today. To give an idea of the abundance of ginseng when white settlers moved in and trade with China began, an average of 140,000 tons of the wild roots were harvested and exported every year throughout the 1770s in North America, and by 1824, a climax of 600,000 tons of exported roots was reached.[9]

Due to overharvesting, land clearing, and change in soil composition (from earthworms, toxins, excess nitrogen), many native species have reached the tipping point and cannot cope as the weedy invasive plants do. Ignorance on the part of wild harvesters in the early days (seeing abundance, many of them thought of profit without consideration of plant depletion) compounded by improper collecting practices has led to near extinction for many species of plants (and animals, for that matter). Something like a "rape and pillage" method was in vogue across the spectrum of species, and this, in part, set in motion the tide of our present situation. Overharvesting of our native medicinal resources has been yet another human disturbance of the native landscape, and it has virtually eliminated many native species of medicinal plants and opened the way for others to move in and replenish the land. Fortunately many of the intruding plants provide powerful remedies to replace the endangered individuals that have been overharvested for medicine, disturbed by development, and poisoned with industrial progress. There is Siberian elm as a substitute for slippery elm, barberry for goldenseal, and purple loosestrife for eyebright.

In the studies often cited to support the undermining of species diversity by weedy invaders, the research is, "in many cases, anecdotal, speculative, or based upon limited field observation."[10] Some researchers cite various means to extinction, with invasive species being one of them, without detailing the specific damages imposed by each. For example, one study includes four other stressors investigated as potential causes—habitat destruction, overharvesting, pollution, and disease—but the authors did not "try to distinguish between major and minor threats to each species because such information was not consistently available."[11] According to Mark Sagoff of the University of Maryland, invasive plants "are generally not more significant contributors to extinction than are native species, off-road vehicles, hunting, weather, fire, contingent events, pesticides, pollution, and many other factors."[12] In a few isolated island and lake ecosystems, there have been some cases of plant extinction, but this hasn't occurred on large landmasses or in ocean environments. According to Sagoff's article,

> [I]n a scientific sample of wetlands, "Exotic species were no more likely to dominate a wetland than native species, and the proportion of dominant exotic species that had a significant negative effect on the native plant community was the same as the proportion of native species with a significant negative effect." In addition, "there was no evidence to support the hypothesis that exotic species are more able to dominate invaded communities because they have fewer natural enemies than native plants"[13]

> *Even the United Nation's Convention on Biological Diversity's Subsidiary Body concluded in their report on invasive species, "there are no records of global extinction of a continental species as a result of invasive species. . . ."*
> DANA JOEL GATTUSO, "INVASIVE SPECIES: ANIMAL, VEGETABLE OR POLITICAL?" THE NATIONAL CENTER FOR PUBLIC POLICY RESEARCH

In fact, quite contrary to predominant invasive species ideology, there are numerous findings describing the actual enhancement of biodiversity in ecosystems with invasive plants. In an article in the journal *Science,* evolutionary biologist Gereet Vermeij writes, "Invasion usually results in the enrichment of biotas [plant and animal life of a particular region] of continents and oceans. In some biotas . . . interchange has pushed diversity to levels higher than the pre-extinction number of species."[14] Fossil records of invasive species movement during the collision of the continents and the surging and shrinking of sea levels reinforce Vermeij's claim; exotics increase biodiversity throughout their new range and fulfill a unique role within the biosystem.

The overall pattern almost always is that there's some net increase in diversity, that seems to be because these [native] communities of species don't completely fill all the niches. The exotics can fit in there.

DR. JAMES BROWN, ECOLOGIST AT THE UNIVERSITY OF NEW MEXICO, IN "FRIENDLY INVADERS"

Species richness is a common means ecologists use to determine the health of a particular ecosystem. Greater species richness in the ecosystem means enhanced productivity and functioning, which in turn leads to the conclusion that by enhancing biodiversity, invasive exotic plants create healthier and more efficient ecosystems.

In "Can Weeds Help Solve the Climate Crisis?" a June 2008 article in the *New York Times,* author Tom Christopher discovers instances where invasive plants adapted to and enhanced ecosystems. He cites biologist Andrew MacDougall of the University of Guelph, Ontario, who was researching the impact of invasive grasses in a rare, biodiverse oak savanna protected by the Nature Conservancy on Vancouver Island in British Columbia. As the rare, native grasses and flowers declined, MacDougall believed the spreading invasives impinged the growth of the native plants, so he hypothesized that removing the exotics would help the recovery of the native flora. But his findings surprised him.

After disproving his own theory, MacDougall began referring to these outsiders as "passengers" in the changing ecosystem, rather than as the "drivers" of this change. He set up different test plots where he removed the weeds by mowing, burning, or by hand—still, there was no significant revival of the native plants. In some cases there was an acceleration of the loss of native flora, and the grassy savanna began to give way to a new environment with woody shrubs and trees spreading throughout. The invasive grasses were in fact providing a *stabilizing element* within the ecosystem and filling in the gaps where the native grasses had disappeared. They stifled the growth of trees and shrubs, preserving the open character of this precious relic of land. MacDougall came to believe that the main reason for the failure of the native species was due to human intervention—*or due to a lack of intervention.* Prior to European settlement, fires cleansed the land periodically (through the practices of native people inhabiting the area or by wildfires), removing plant matter and reviving the soil. In the absence of regular fires, the organic material built up, and the native flora failed to thrive in such an environment.

MacDougall also believes climatic changes are partly responsible for the success of these invasive plants. He notes, "Weather records reveal that spring warmth in this semiarid region is coming earlier than it used to, and the season's rain is more consistent. The wheatgrass, which awakens from winter dormancy earlier than the native grass species, has gained a competitive advantage from this change."[15]

> *All living beings have the right to engage in the struggle for existence.*
>
> L. H. BAILEY

Most researchers, including myself, have found it incredibly difficult to find any proven cases of extinction due to plant invasions. Many scholars echo this lack of credibility. Mark Sagoff comments, "I have yet to find evidence that invasive plants are the main cause of extinction except for certain predators in lakes and other small island-like environments," and it is hard to find a single example of an extinction anywhere caused principally by introduced plants. "In fact, there are surprisingly few instances in which extinctions of resident species can be attributed to competition from new species."[16] Davis adds that "there is no evidence that even a single long-term resident species has been driven to extinction, or even extirpated within a single US state, due to competition from an introduced plant species."[17] A study of U.S. Fish and Wildlife Service data identified the primary causes of endangerment for ninety-eight plant species protected under the Endangered Species Act. The authors "reported that invasive species posed no more of a threat than off-road vehicles to these 98 endangered plants."[18] Vermeij writes, "The evidence so far points to the conclusion that invaders often cause extinction on oceanic islands and in lakes, but rarely in the sea or on large land masses."[19]

Yet even in island environments such as on Hawai'i, where the native flora consists of about 1,100 species, an additional 4,600 exotic plants have been identified with about ninety reported extinctions since the arrival of Europeans. Despite the low ratio of native plant disappearances to exotic influx, Hawai'i has become a poster child for loss of plant species in the United States. Hawai'i also might be the ecosystem of greatest flux, with the environment on the whole forming and melting islands all the time. On the rich island landscape of New Zealand, 2,065 unique native species now intermingle with 22,000 new plant species, 2,069 of which have naturalized since colonialization. With this great influx of foreign flora, the island has documented just three plant extinctions. At least 4,000 plant species have become naturalized and established to various degrees in North America during the past five hundred years, and these exotics now represent nearly 20 percent of the continent's plant species. According to the evidence, invasions by foreign plants increase the richness of biodiversity within the greater ecosystems.[20]

> *Wait, you say, what about those plants that deprive the native wildlife and those that poison the soil, killing other plants?*

In fact, some surprising interchanges between invasive plants and other species show the benefits of the invasives: how they meld with the ecosystem and provide for its inhabitants—contrary to the stigma that they're ruthlessly hostile to native insects and wildlife. Cases abound relating the adaptation of native species to foreign individuals. Microbes, insects, birds, and mammals learn how to make the best use of the new plants. Just as native flowers do, a wide range of foreign flowers continue to provide plentiful nectar and pollen for winged invertebrates, and the berries and shelter provided by these new plants are beneficial to much of the wildlife, and even essential to the survival of birds such as the endangered southwestern willow flycatcher (*Empidonax trailii extimus*). This bird relies on salt cedar for shelter from the desert sun.

Studies have shown that invasives do not outcompete native flora for insects and that "native plants can also benefit from the invasive ones" by increasing overall insect visits.[21] A USDA study found that spotted knapweed affects the habits of predator species in their ecosystem (in this case, spiders). In the invaded areas, the spiders were collectively thirty-eight times more abundant, created more expansive webs that caught increased numbers of larger prey, and were twice as likely to reproduce on spotted knapweed land than in uninvaded grasslands.[22] Some people have deemed as useless the wetlands densely covered with plants such as phragmites, though upon closer examination these plants provide rich sources of food to many life-forms. In most field studies phragmites coexists in marshes with diverse and abundant biota, and it provides food for insects, birds, and microscopic organisms much as the native cordgrass marshes do. Contrary to common projections about these plants, these findings uncover just some of the unexpected impacts of the opportunistic plants that can revive the wild.

One of the other criticisms of plant invaders is their purported attack on other species of plants with their allelochemicals—the chemicals sent into the surrounding soil to inhibit the growth of others nearby. Many species of plants and trees do this—including American-born black walnut, with the compound juglone, and various native pine trees that acidify the soil. The exotic spotted knapweed (*Centaurea maculosa*) is said to use these "weapons" against the native flora of the western rangelands. In this case the component *catechin* has been said to kill nearby plant neighbors. Yet a pair of USDA studies found that this has been misguided science and that "alternative mechanisms must be found to explain the success of this species as an invader in North America."[23] Instead of proving the repeated claim that catechin causes oxidative stress to the surrounding plant life, this study found it acting as a potent antioxidant to the soil and cleaning it of any free radicals or toxic and imbalanced microbial life.[24] A separate study at the University of Massachusetts, Amherst, reveals that other phytotoxins of diffuse knapweed (*Centaurea diffusa*) facilitates nutrient uptake for the plant, helping it acquire iron, which is scarce in many of the deprived, alkaline-rich environments it colonizes.[25] Many other invasive plants, such as Russian olive, Scotch broom, and dandelion, help accumulate nitrogen and other trace elements to enrich depleted soils, which can also lessen the need for chemical fertilizers on agricultural lands. These plants' ability to provide scarce nutrients—a fact unrecognized by most assertions about invasives—represents another advantage of opportunists over natives, suggesting these invasions are more an expression of the land and benefiting the environment than of competition between plants.

HYBRIDIZED HARMS

Environmental harm is supposedly additionally substantiated by cases of hybridization—that is, invaders reproducing with natives. There is a concern that, with time, hybrids could eliminate native genetic strains. White mulberry (*Morus alba*), originally from Asia and now widespread in eastern North America, often hybridizes with the native red mulberry (*Morus rubra*), which confuses the two and makes it difficult to identify them, though there is no reported loss of *M. rubra*. In nineteenth-century England salt marsh cordgrass was introduced from North America, and in turn, it hybridized with the native small cordgrass, creating a new offspring. Now called common cordgrass, it is a distinct species that is incapable of reproducing with either of the original species, and it has not eliminated the native variety. Today, scientists use the invasive tree Siberian elm to hybridize with the American elm for building resistence to Dutch elm disease. Hybridization is, in fact, a natural process that has long had a hand in evolving plants. About a third of the plant species around us today have been formed in this way.[26] Hybrid adaptations further biodiversity and actually enhance plants' genetics, strengthening their ability to handle environmental stressors and to pass along their chlorophyll bloodline. We must remember that most of our food—along with our ornamental cultivars and nonedible cash crops—is composed of hybridized plants. Many animals close to us reap the benefits of hybrid vigor. Our cattle, chickens, dogs, and cats—and even us humans—have been hybridizing for a long time to increase genetic makeup for higher adaptive qualities and for greater "productivity."

> *This is just what nature does. Yet companies such as Monsanto are allowed to genetically cross strains of plants with nonplant fragments—effectively forcing DNA links to create something of Monsanto's own design. This is not what nature does.*

HARM TO HUMAN HEALTH?

The first thing I want to say as a romantic herbalist: They're medicine, silly!

We'll get to the medicine later, but first we'll deal with cases of minor health issues arising from the handling of a few plants. For example, the reputation of tree-of-heaven (*Ailanthus altissima*) is that it can cause great harm: it can potentially cause heart problems, debilitating headaches, and nausea in people who do not protect themselves from exposure when they cut and handle the trees. Chinese immigrants, however, cherished the tree-of-heaven for its glorified health benefits. In fact, they brought this plant with them from across the ocean.

There must be something people are missing.

Other plants use thorns, thistles, burrs, and irritating chemicals for defense from predators—such as livestock—that can, for instance, disturb overgrazed land. There are also widely dispersed native plants that harm: poison ivy and poison oak (*Toxicodendron* spp.), water hemlock (*Cicuta maculata*), and cow parsnip (*Heracleum lanatum*), which are highly irritating to the skin when touched. When working with these plants, whether for medicine or eradication, wearing protective gear is a commonsense remedy that eliminates the chance of harm in most cases.

BIOLOGICAL CONTROLS

As for any biological control of invasive plants, there is limited understanding of the possible consequences of controlling these plants by releasing pests into a complex ecological matrix. The unpredictable approach of introducing a biological control does not kill all the targeted weeds, and there are unknown impacts of pests from foreign environments: how they relate to the native landscape and what the potential harms might be with their long-term occupation. The intentional unleashing of these pests (insects, fungi, bacteria, and viruses) into the environment in order to kill an unfavored plant only leaves them hungry for other plant species if these biological controls flourish. It is dangerous to risk the potential of creating a "home" for an invasive pest in order to rid the environment of an invasive plant.

Indeed, I cringe when I think that the great American chestnut forests were toppled by a simple fungus. I understand the chestnuts weren't undercut by the intentional release of a pest. The fungus simply hitched a ride, unidentified on some saplings from across the seas. Devastation as a result of using a biological control has happened and is happening—and it amounts to the real potential for another chestnut blight. These pests need only carry with them a disease that is unforeseen or unknown—and then they can set about infecting at will.

> *Frightening headlines might read something like: "Unknown Virus Carried by Beetle to Control Loosestrife Wipes Out All Apple Trees."*

One notorious example in the failed attempt at using a biological control is the cane toad (*Bufo marinus*), which was introduced to Australia in 1935 by the sugar cane industry in hopes of controlling a beetle pest. After thousands of toads were released without substantial ecological assessment, officials found that not only did they fail to address the beetle problem, but also the toads began eating anything that moved— and enjoyed an unlimited check by any predator species and continue to wreak havoc on Australia's unique biosystem and inhabitants. Another report, this one from *Science,* illustrates the uncontrollable nature of an insect used to control exotic thistles. Within a four-year period the weevil (*Rhinocyllus conicus* Froeh.) greatly expanded its territory and developed a preference for the native thistles that grew in the same area, and ultimately the weevils significantly reduced the native thistle's seed production.[27] We don't fully comprehend the long-term ecological effects of biological controls in the wild, and therefore we cannot properly evaluate or regulate these introductions in the future.

As we have seen, then, there are numerous players in the invasive scenario and great repercussions in trying to control this dilemma. The health of an environment comprises various species, and isolating any one plant leads to misunderstanding complex processes and their interrelationships. Likewise, harm to an environment is not easily defined and observed, for worlds of invisible elements are unobservable by our standard vision.

> *An imaginary conversation about a real happening on a windswept Australian island:*
>
> *"We have a problem here. These exotic feral cats are eating the native birds."*
>
> *"Let's get rid of the cats!"*
>
> *"We have the birds back, but there's another problem. Without the cats, the exotic rabbits are now out of control and are eating all the native plants."*
>
> *"Let's get rid of the rabbits!"*
>
> *"We have the native plants back, but without the rabbits to keep this exotic plant in check, it has been spread all over the island by the birds eating and pooping everywhere . . ."*
>
> *On and on and on it goes. Pull on one thread, and the whole garment falls apart.*

5
Invasive Herbicidal Impacts

Man can hardly even recognize the devils of his own creation.

ALBERT SCHWEITZER

WHEN CONSIDERING THE ENVIRONMENTAL harm attributed to invasive plants, we must assess the impact of herbicidal eradication efforts. As the years go on, millions of gallons of toxic herbicides are sprayed throughout the environment, contaminating the land, air, and waters from which we eat, breathe, and drink. Alarm over widespread devastation from herbicides and pesticides has been sounding for more than forty-five years, ever since Rachel Carson, in her groundbreaking treatise, *Silent Spring,* spelled out the dangers of these chemicals—but we have all become conditioned to mute this bell that constantly rings. We have learned to banish these worries from our minds, not realizing the importance of Carson's original message that has since been retold innumerable times. Three generations have been born since Carson's publication—and thousands of chemicals have leached into our bodies, many even before our first breath. Yet we continually forsake our families and communities in the name of progress.

The torturous, slave-driven human fallacy of progress has led to much of our history's demise and darkness . . . and now we stand at the brink of regress.

Herbicides are the most common means to eradicate targeted plants, and in some cases they are said to be the only option in extreme invasions. The hazards of these chemicals are widely known and are often printed clearly on the label. Study after study reveals their deep impacts on all the earth's creatures, great and small. Herbicides damage an organism's neurological, reproductive, endocrine, metabolic, and immune systems, causing genetic damage, birth defects, and various cancers. Children and pets are particularly vulnerable to the hazards of herbicides, because they are often in close contact with lawns and the outdoors, and children's delicately growing bodies are susceptible to the minutest concentrations of these chemicals. Even if the poisons are carefully applied (and they aren't most of the time), they eventually contaminate the water, soil, and air and enter the food chain, affecting microorganisms up through to our dinner plates.

The outcome: millions of people and countless life-forms are unintentionally poisoned every year by pesticides. In a U.S. Geological Survey, more than 60 percent of the air tested was found to contain the common herbicide Tordon (picloram and 2, 4-D). Further, an ounce of some pesticides can contaminate seventy-eight hundred gallons of groundwater. Upward of 85 percent of the chemicals used today in the field do not have proper safety information, and some demonstrate extremely strong impacts at very low levels of concentration, whereas at the same time herbicide use has grown by 383 million pounds from 1996 to 2008, with 46 percent of the total increase occurring in 2007 and 2008.[1] Though study after study reveals the hazards of these chemicals (and some government agencies compensate victims and some consumers win lawsuits against pesticide manufacturers), bureaucracy continues to allow these toxins freely to enter and alter ecosystems and thereby to negatively impact our health.

Herbicides (a.k.a. plant poisons) are indiscriminate in their cellular tastes.

AGENTS OF WAR: AGENT ORANGE AND THE RAINBOW HERBICIDES

The use of herbicides for warfare was first brought to our attention in the Vietnam War, when *rainbow herbicides* were sprayed across territories to reveal hideouts, destroy agriculture, and poison the enemy. The barrels containing these agents that Dow Chemical Company and Monsanto, among others, manufactured had a colored stripe painted on them to identify the contents:

> *Agent Orange, Agent Green, Agent Pink, Agent White, and Agent Purple*

The most common was Agent Orange, an equal blend of two phenoxy herbicides (2, 4-D and 2, 4, 5-T). Between 1961 and 1971, about forty-six thousand tons of it was sprayed at intensified rates over 3.5 million acres of southern Vietnamese forests and cropland. Not only were ecosystems completely ravaged by this mass poisoning effort, but also millions of civilians and allied troops were caught in the crossfire. The toxin dioxin used in all of these poisons has been reported by the U.S. Department of Veterans Affairs to cause a wide variety of illnesses that affect various bodily systems and is still present in our environment at high concentrations. Some known ailments that are compensated under VA benefits include type 2 diabetes, prostate cancer, respiratory cancers, multiple myeloma, Hodgkins disease, non-Hodgkins lymphoma, soft-tissue sarcoma, chloracne, porphyries cutanea tarda, peripheral neuropathy, and spina bifida in the children of veterans. Since 1984, Dow Chemical Company has lost various class-action lawsuits regarding these poisonings of American, Australian, Canadian, New Zealand, and South Korean veterans in Vietnam. All have won health care compensation for the unforeseen hazards of their service.[2]

ROUNDUP: THE MIRACLE WEED KILLER

"Safe as table salt," proponents claim. . . . Well, I want to see you sprinkle it on your food.

Among other products, Monsanto Corporation has been an inventer and manufacturer of genetically engineered crops, bovine growth hormone, dioxin, PCBs, TNT, Agent Orange, and the infamous Roundup weed killer. Monsanto's profits for 2008 neared two billion dollars for their line of glyphosate-based herbicides—a figure nearly equaling their corn seed business profits. These types of broad-spectrum herbicides became prominent in the 1990s, when Monsanto and other large corporations began inserting glyphosate-resistant genes into various crops. They can now be found in corn, cotton, soybeans, rapeseed (the origin of canola oil), and sugarbeets, all common adulterated plants that can withstand massive amounts of herbicide. (Interestingly, soy and canola oil in our food chain are 90 percent genetically modified.) Touted as safe and environmentally friendly, hundreds of thousands of tons of glyphosate are spread across farmland throughout the world every year, and permits are being granted all the time to allow increased herbicidal residues on food crops. At one point, the FDA allowed Monsanto to increase residues from six parts per million (ppm) to twenty ppm on their genetically modified (GMO) soybeans. Due to loosening regulations and increasing circulation of the GMO crops that tolerate higher amounts of herbicides, we humans are constantly inundated with toxic food and plants that continue to flourish in these environments. The Union of Concerned Scientists reports that these genetically engineered crops actually help weedy plants to develop greater adaptive traits—thereby ensuring that invasives are stronger, bigger, and "scarier." Though Monsanto's original intent in genetically modifying crops was to increase production, the GMO corn and soybeans designed to resist insects and tolerate glyphosate have not proved to fulfill their promise of greater yields, and the weedy plants continue to outcompete them for nutrients and water.

Though Roundup has been touted by Monsanto right on the bottle as safe and environmentally friendly, and has been backed by complacent U.S. government regulators, much evidence proves otherwise. The state of New York sued Monsanto for its safety claims, resulting in a cease and desist order regarding the use of those terms, along with a two-hundred-fifty-thousand-dollar payout settlement, even though, all the while, Monsanto claimed no wrongdoing.

The ingredients in Roundup include: isopropylamine salt of glyphosate (active ingredient), water, ethoxylated tallowamine surfactant, related organic acids of glyphosate, and excess isopropylamine. The following inert ingredients are in Roundup Super Concentrate Weed and Grass Killer: polyoxyethylene alkylamine, water, and FD&C Blue No. 1.

Research completed by Monsanto's own toxicologists regarding Roundup's toxicity has shown that ingestion causes "irritation of the oral mucous membrane and gastrointestinal tract . . . pulmonary dysfunction, oliguria (scanty urination), metabolic acidosis, hypotension, leukocytosis and fever."[3] Anecdotal evidence reports a wide range of central nervous system damage, and a 1980 Environmental Protection Agency study showed that ninety-four people exposed to glyphosate reported these symptoms: bronchial constriction; pleuritic chest pain and nasal congestion; blurred vision, corneal erosion, and conjunctivitis; contact dermatitis; headache; nausea, diarrhea, and abdominal pain; irritability; excessive sweating; vertigo; malaise; swelling of the extremities; and nervous system disorders.[4]

Another study by David H. Monroe, an Industrial and Environmental Toxicologist, performed on another Monsanto product revealed that it contained 1,4-dioxane, a contaminant in the surfactants used. This toxin was present at a level of three hundred fifty ppm. It is a known carcinogen and damages the liver, kidneys, brain, and lungs—but its producer and the EPA continue to deem it a safe product.[5]

HERBICIDAL CASTRATION

Atrazine is one of the most widely used herbicides in the world, and it is a common applicant to corn crops across America. Since the 1990s, reports of its devastating health effects have been released. The European Union banned the chemical in 2005, and some states in the United States are also mulling plans to ban the chemical. As in other toxic pesticide cases, class action lawsuits have been brought to court to stop the use of this poisonous chemical, though the EPA has continually turned a blind eye and has allowed widespread concentrations of atrazine to accumulate over the years.

Studies have shown horrifying effects on frogs exposed to atrazine at levels considered safe by the EPA. One study illustrated that three-quarters of male frogs were chemically castrated after exposure to atrazine and were therefore considered dead—that is, they had no ability to reproduce in the wild. An additional 10 percent of the male frogs actually turned into female frogs! Adaptations caused by the chemical gave these frogs the ability to make eggs and reproduce with other male frogs—though they produced only male offspring. An additional study showed that atrazine alters frog tadpoles into hermaphrodites (that is, they possess both male and female sex organs) at concentration levels as low as 0.1 parts per billion (ppb)—which is thirty times lower than levels allowed in drinking water by the EPA (3 ppb).[6]

The manufacturers of atrazine, including Syngenta and Monsanto, dispute the claims of these studies and others, and the EPA is slow to respond to these findings that show detrimental health effects. Reseacher Tyrone B. Hayes at the University of California at Berkeley argues against these injustices: "When you have studies all over the world showing problems with atrazine in every vertebrate that has been looked at—fish, frogs, reptiles, birds, mammals—all of them can't be wrong."[7] Atrazine is one more example of fraudulent chemical companies and subserviant government policies allowing toxic herbicides to silently poison our surrounds and our bodies, negatively impacting future generations of all life on Earth.

FURTHER ENVIRONMENTAL IMPACTS OF HERBICIDES

Yet, the harder they hit us,
the louder we become.
Kind of like
the skin on a drum.

MICHAEL FRANTI, SONGWRITER-ACTIVIST,
"SKIN ON THE DRUM"

In addition to the harmful health impacts of herbicides, the further environmental influences on plants themselves are equally as horrifying. A common problem on rangelands when some plants are subjected to spraying is that the herbicides can actually increase toxins within the plant (including potassium nitrate and cyanide), thereby poisoning the grazing livestock and wildlife that consume it. Herbicides try not only to kill the target weed, but also they will often destroy most other wild species. These poisons then further permeate the environment, polluting the soil and water and continuing to inhibit most plant growth except for the plants that can tolerate it. This sometimes leads to land that is barren of any plant life except for the species intended to be killed. Most herbicide-tolerant plants are the weeds that we are trying to suppress, and the use of Roundup in agriculture is creating an ever-growing list of plants who are resistant to glyphosate. At this point in history at least nine species, including Palmer amaranth (*Amaranthus palmeri*) and giant ragweed (*Ambrosia trifida*), are as unaffected by sprayings as the GMO plants designed to withstand increasing loads of poison. Even with tolerance on the rise, companies such as Bayer and Dow AgroSciences are brandishing new lines of GMO seeds to withstand the feedings of their own glufosinate and glyphosate herbicides.

> *The rapid adoption of glyphosate-tolerant traits in recent years has spurred a growing issue with resistance to that herbicide in key weeds. Our new family of traits will significantly expand the herbicide toolbox, ensuring that weeds have a lot less opportunity to build up their resistance.*
>
> JEROME PERIBERE, DOW AGROSCIENCES PRESIDENT
> AND CEO, "DOW AGROSCIENCES UNVEILS
> BREAKTHROUGH IN MULTIPLE HERBICIDE
> TOLERANCE TRAITS" (DOW NEWS REALEASE)

Here we go again . . .
The eco-warriors enthusiastically raise their arms,

and are called to battle with honorable conviction,
not knowing they are actually creating more harm than good,
by their blindness of intentions.

Some Alternatives to Herbicides

pulling

cutting

digging

mowing

livestocking

mulching

planting

feeling

thinking

smelling

drinking

eating

listening

loving

thanking

6

The Economics of Weeds

These are economic problems for humanity, not problems for ecosystems.

DAVID THEODOROPOULOS, *INVASION BIOLOGY: CRITIQUE OF A PSEUDOSCIENCE*

THE ECONOMIC LOSSES DUE to some opportunistic plants are the most viable arguments against invasives presented by the agricultural industry via the herbicide companies and sometimes industries such as tourism and commercial or recreational fishing. Every year the typical industrial farmer invests much money in preparation for the growing season, and he or she risks potential loss and failure of crops and grazing land. There is no insurance against the potential financial risk, and often a season could result in a loss quantified by varying numbers. In addition, modern-day farmers shoulder the rising costs of being bound to corporations who force them to buy GMO seeds and the appropriate herbicide every year—meaning they cannot grow the seeds from the previous year's harvest. The most substantial losses imposed upon the farmer by opportunistic plants are related to industrialized agriculture that relies on monocrops, and the weeds cost U.S. farmers about 12 percent of their harvest, estimating an annual loss of thirty-three billion dollars. The National Forest Foundation estimates that each year the United States spends thirteen billion dollars in fighting noxious plants; Cornell University reports $34.7 billion spent annually on combating the effects of invasives on the nation's agriculture, water quality, wildlife, and recreation; and a much-cited Pimentel paper reporting on all fifty thousand nonindigenous species (plant, animal, insect, and microbe) claims environmental damage and losses that total approximately one hundred thirty-seven billion dollars per year! (One example in the report: thirty dollars for each songbird killed by the exotic cat amounts to thirty dollars lost—and annual losses attributed to pet and feral felines total seventeen billion dollars [approximately one-eighth of the total]!)[1] Yet, even with these substantiated claims, there is still controversy: the Congressional Research Service conducting a report in 1999 stated that "while the damage from some non-native species can be great, few have proved to be economically harmful, and many are beneficial."[2] These widely circulated and often-exaggerated economic claims continue the deception of a great catastrophe, which benefits a select group of industries while offering no benefit to the people and the planet as a whole.

So who is benefiting from the War on Plants?

FOLLOW THE MONEY

Nazi Germany pioneered chemical engineering for combating plants, pests, and people by developing highly poisonous organophosphate compounds used in agricultural pesticides and as chemical warfare nerve gases. In America after the two World Wars were over, there was a movement to find use for the millions of pounds of wasted ammunition and explosives that remained. Factories that once manufactured war machinery were waiting to be filled, soldiers needed jobs, and there were plenty of raw materials to use. The first widely used herbicides and pesticides were nothing but leftover weapons of war. Nitrogen- and phosphorus-based compounds accumulated in massive, stockpiled amounts during wartime, which then led to the practice of discarding them on agricultural fields as a synthetic fertilizer throughout America and, eventually, the world.

DuPont was the largest manufacturer of gunpowder during WWI and now is the parent company of the world's largest seed company, Pioneer HiBred, and Monsanto saw a one-hundred-fold increase in profits by supplying chemicals to produce highly reactive explosives such as

TNT. Dow Chemical and Monsanto have been the leading manufacturers of herbicides for decades, reaping huge profits from Agent Orange's campaign against the Vietnamese jungles and with the Roundup family of herbicides for every dangerous plant imaginable.[3]

> *These guys keep popping up . . . like weeds on the page.*

Thus the promise to farmers of abundant harvests and increased profits by flushing nutrients and eliminating pests—and the promise of creating an unthinkable thirty acres of corn in the 1940s. Not until the introduction of herbicides, synthetic fertilizers, and tractors could larger-scale farming take place by an individual: how could the average farmer at the time not want to jump on board? Joe Farmer found that he could in the short term make more money with less work. He likely thought that spraying the weeds sure beat the hand and hoe.

> *The temptation was great and foresight lacking, with some on the wrong side of the fence.*

Giants such as Monsanto now control the seeds used by large-scale farms and provide the appropriate herbicide in which farmers can cover their crops. No doubt the heads of Monsanto laugh at the madness they have created and relish the billion dollar profits they garner from farmers with indentured repeat business. The flourishing agrichemical industry capitalizes on farmers who spend sometimes one hundred dollars per acre and at least ten billion dollars in herbicides annually worldwide. Meanwhile, most opportunist plants are growing resistant to these poisons, and all the time new, unknown chemical formulations are being made for combat. Ironically, plants are actually learning to thrive in these toxic environments, and it is mainly large-scale, nondiversified factory farms that complain of losses. Smaller, diversified organic farms report far fewer issues because they do not use chemical pesticides.

Once the territory under human control (farmland) was in the grip of the chemical industry, a new terrain was mapped. The new plant enemy became those outside the fence-lined agricultural fields, and new ecological soldiers were recruited for combat to rapidly respond against these plant outsiders. When certain camps were infiltrated by chemical stewards, the alarm became louder and more prevalent. The focus was now on the wild places where native species teetered on the brink. We hear the same sirens now—but from new sources. Most nature-based organizations and tourism offices now head to the trenches, convinced it is better for the environment to spread herbicides and distrupt the ecosystem than to let a natural process play out. Regulatory committees and invasive plant councils have been overwhelmed by herbicide company representatives, thus the message has become widespread—which ensures high profits for the herbicide corporations due to high sales and repeat business. The economic costs seem to benefit a select group of commercial businesses and industries, and large chemical companies financially assist various research studies and programs for universities, environmental groups, and nature organizations that pledge to use and promote the use of herbicides.

Chemical companies spread their agenda by supporting various invasive plant coalitions—universities, environmental conservation groups, state and federal agencies—by spending millions of dollars in grants to further manipulate the public's image of both the benevolence of the companies themselves as well as the malignant nature of certain green terrestrials. The multinational chemical industry companies BASF, Dow Chemical, Monsanto, Bayer, and Syngenta have all provided millions of dollars in grants to assist groups over the years, but in order to receive these funds, these recipient organizations must tout the benefits of herbicides. Dependent on these valuable funds, academia misleads science, environmental groups turn to these toxic solutions, and the government provides the policies that continue to plunder our taxpayers' money and toxify our environment—then, to close the circle, it applies for more funds. Favorable reports are published, articles are written, the EPA turns a blind eye, and documentaries are made with the invisible hand of profiteering chemical companies. One such documentary, *Silent Invaders: Plants out of Place II,* was made possible through educational grants from BASF Vegetation Management Group and Syngenta Professional Products in collaboration with various U.S. government agencies.

Well-meaning conservation groups such as the Nature Conservancy, the National Wildlife Refuge Association (NWRA), the Sierra Club, and others have all taken the aggressive stance of supporting ridding the natural environment of opportunistic wild plants by using toxic herbicides and high-impact removal methods that further disrupt areas (which is exactly what invasives like us to do).

> *If you ask me, these are strange policies for the promoters of a wild and changing wilderness.*

The Nature Conservancy has introduced SWAT (Strategic Weed Action Teams) to combat invasions, and the organization's task force recommended that the nonprofit "elevate the political profile of the invasive alien species issue to establish new funding and policy support,"[4] even while a study on their own property has proved that these so-called dangerous plants have actually benefited and helped to preserve their endangered landscape. In 2002 the NWRA urged congress to spend one hundred fifty million dollars to "train and mobilize 5000 volunteers . . . deploy 50 rapid response 'strike teams' to combat early infestations . . . [and] implement the Management Plan of the National Invasive Species Council."[5] Relying on volunteers, the NWRA has estimated that their savings annually from thirty-six thousand people working more than 1.3 million hours without pay to be fourteen million dollars. The Sierra Club conducts community workshops and gets paid by people (five hundred dollars per person) to take them on invasive plants–eradication outings. These nature-based organizations have overshot their mark on this dilemma—and have been forced to rely on this mentality to further fund their causes. They continue to provide the authority and so-called natural approach to these invasive species by supporting the further disturbance of precious habitats with poisons, high-impact removal policies, and a hate mentality, and they continue to distract the masses from the main source of these problems.

> *Hollywood will be on board soon. I can see the movie now:* Attack of the Killer Loosestrife.
>
> *Why would some choose to create and spread these exaggerated claims? Follow the money. . . . There are always some without ethics who will disseminate propaganda in*

order to climb ladders and make cash. Chemical producers, pharmaceutical companies, and many of our governing officials have, throughout history, repeatedly abandoned their morals and discredited obvious atrocities in order to further their own agenda, make a profit, and seize more power.

Prop·a·gan·da *n. 1. Roman Catholic Church; a committee of cardinals, the Congregation for the Propagation of the Faith, in charge of the foreign missions 2. Any systematic, widespread dissemination or promotion of particular ideas, doctrines, practices, etc., to further one's own cause or to damage an opposing one 3. Ideas, doctrines, or allegations so spread: now often used disparagingly to connote deception or distortion.*

WEBSTER'S NEW WORLD DICTIONARY

An unruly invasive idea planted by exotic missionaries in the native mind of men:

> *Propagules on the foreign landscape,*
> *filling in the holes they create,*
> *helping the dissemination of their fate,*
> *by seemingly to complicate.*

How much do we value the nectar they produce for bees and butterflies? How about a price per pound for the berries they grow that feed the birds and other wildlife? How much do we value their medicines, from which pharmaceutical companies make millions of dollars? How about air-, water-, and soil-cleansing plants, which ask for no fees to do our dirty work? What is the potential net income for these economically beneficial and prolific plants?

ECONOMIC BENEFITS OF INVASIVE PLANTS

Nature abounds with vast resources, but it takes the inventiveness of the human mind to create ways in which to transform the vitality within these seemingly inert sources. The sun, waves, wind, rivers, trees, earth, and even our own waste have been transformed into valuable commodities that can, in turn, power so much. Invasive plants, with the vast acreage they cover and their biomass, can be used creatively and can be transformed in new ways that we have not yet even perceived.

> *First of all, money need not be spent on costly and deadly herbicides. Lower expenses and starve the Beast.*

Food and Fodder

Exotic invasive plants already provide substantial food to the inhabitants of their environments, and some provide direct profit for industry. Some of the most prolific spreading plants, such as purple loosestrife, yellow star thistle, and kudzu, serve as valuable nectaries for the exotic honeybee, helping to sustain a struggling bee population and apiary industry. Honey and bee pollen are a multimillion dollar source of revenue that is overlooked by the economic critics of invasive plants. There is growing evidence that herbicides used on many invasive plants visited by bees could be contributing to collapsing bee populations. German regulators banned chlothianidine and related chemicals in 2008 after this family of pesticides was held responsible for the death of about eleven thousand bee colonies.[6]

In recent history, many foreign plants were promoted and used as fodder for farm animals and for other supplementing agricultural purposes. Kudzu is a valuable commodity in its native Japan; the nutritious root starch is used as a thickening agent and flour, and its leaves and flowers are eaten by humans and livestock. Most invasives also serve a wide variety of other wildlife, providing food and shelter. Barberry, Russian olive, and blackberry can sustain wildlife with their fruits, and salt cedar shelters the endangered southwestern willow flycatcher. Ultimately, with the complex tapestry of interdependent plant, animal, insect, and microbe life, every plant somehow provides for or feeds another species.

An exciting emergence in response to the invasive species crisis has been the creative use of these plants as food. *The Invasive Species Cookbook* by J. M. Franke is amusingly subtitled *Conservation through Gastronomy,* and some communities now promote invasive species cook-offs to explore overabundance problems with practical solutions.

> On the menu tonight:
>> *Garlic Mustard Pesto Pasta*
>> *Dandelion and Lambs-quarters Greek Salad*
>> *Kudzu Leaves Stuffed with Wild Boar*
>> *Roasted Thistle Heart*
>
> And to drink:
>> *Dandelion Wine*
>> *Honeysuckle and Rose Hip Tea*
>> *Dandelion Root Coffee*
>
> For Dessert:
>> *Strawberry Knotweed Pie*
>> *Autumn Olive–Blackberry Tart*

Many of our desolate and poor lands, where these plants thrive, are also home to economically impoverished communities, and therefore they could lead to opportunities: citizens could harvest them and nourish themselves for free.

Biofuels

In a world in need of renewable fuel sources, these overabundant plants make an ideal opportunity for use as biomass-biofuel energy sources. According to some, kudzu is a perfect biofuel. Its roots contain as much as 50 percent starch by weight—an ideal composition for ethanol production—and the vines grow a foot a day, thus they represent plentiful biomass energy waiting to be harnessed. Even an article in the racing portion of ESPN suggested kudzu is "NASCAR's fuel of the future."[7] Switchgrass, a problem plant in the Midwest, is now being harvested for high-quality, energy-efficient, clean-burning biofuel. Switchgrass ethanol can yield 540 percent more energy than it requires in its production, and burning it releases 94 percent less greenhouse gas than gasoline. Contrast this with the use of corn ethanol: corn releases only 22 percent less greenhouse gas than gasoline and is also an inefficient use of the plant. (Corn used as biofuel incorporates only the kernels instead of the whole plant and also diverts a much-needed food source.)[8] Japanese knot-weed grows at least two inches per day, creating high biomass and leading to two to three harvests per year. What's more, these plants grow abundantly along roadsides. By using machinery already available, with some slight altera-tions, harvesting would be easy and convenient. Many other opportunistic plants have high seed counts, meaning we might extract from them volumes of oil, which can be made into fuel, food, paint, varnish, cosmetics, and so forth—just as were used prior to the petrochemical revolution.

Medicines

Various components in plants have been extracted into medicines reaping hundreds of millions of dollars for pharmaceutical companies. In April 2008, the pharmaceutical company Sertris was bought out by pharma-ceutical giant GlaxoSmithKline for seven hundred twenty million dollars. GlaxoSmithKline wanted to cash in on the profits of resveratrol, the won-der drug found in red grapes and Japanese knotweed. In fact, Japanese knot-weed has the greatest concentration of resveratrol of any plant in the world, and most companies are using this plant to extract this compound. The cost of a bottle of tablets (a clump of roots) of resveratrol in the store is twenty to thirty dollars. There are high profit potentials in this wild plant that grows everywhere. The weedy Saint-John's-wort makes at least five hundred million dollars annually for the herbal industry, and dandelion is one of the most common herbs in basic tea blends and coffee substitutes. We find many of these plants among the pages of the Chinese Materia Medica; they have documented historical uses that have been proved effective over thou-sands of years. Kudzu, honeysuckle, dodder, knotweed, and tree-of-heaven are all very much utilized for various conditions. Barberry is a perfect substi-tute for the noted herbal antibiotics Chinese coptis and goldenseal, contain-ing *berberine,* an isolated drug used in China for more than fifty years to treat a wide range of infections. The potent antimalarial agent *artemisinin* is derived from the endemic plant sweet Annie (*Artemisia annua*) and has been made into a concentrated drug to treat malaria in infested areas. An abundance of other phytochemicals have been further refined into pharma-ceuticals and herbal supplements in order to treat a variety of conditions and infections, and many are present in invasive plants. In chapter 9 and in part 3 of this book we will further explore a medicinal system based on these widespread plants.

> *Given that effective medicinal plant extracts could shift the benefit:cost ratio from dollars to pennies, and that many known antimalarial plants, including* A. annua, *grow prolifically . . . this could significantly change the societal and economic burden of disease in many parts of the world. In addition, properly planned cottage industries producing plant-based remedies for the treatment of malaria and other disorders could generate income for rural communities.*
>
> KEVIN SPELMAN, PH.D., "'SILVER BULLET'
> DRUGS VS. TRADITIONAL HERBAL REMEDIES:
> PERSPECTIVES ON MALARIA"

Phytoremediation

Most of the plants described as villainous, low-life burdens who wreak economic havoc generally provide the environment with services that otherwise would come with enormous price tags. Hundreds of billions of dollars are spent cleaning known polluted land and waters throughout the United States, yet the invasive plants that serve as water filters (common reed, water hyacinth, purple loosestrife), soil cleansers (Japanese knotweed, dandelion, knapweed), soil stabilizers and protectors (kudzu, blackberry, barberry), soil enhancers (Russian olive, Scotch broom), and air purifiers (English ivy, tree-of-heaven) provide a free service of great importance, and the financial gains of their service are never reported.

Great efforts to eradicate exotic species continue to despite the shining economic and environmental benefits of these plants. A nonplant example of a twofold environmental and economic aid is the zebra mussel, a notorious invasive species that has proliferated throughout the Great Lakes. While extensive efforts to remove them were under way and billions of taxpayer's dollars were spent, the creature actually cleaned these bodies of water. The most noticeable effects were in Lake Erie, which for decades had been a toxic dump for industries and was full of pollutants and algae. The U.S. Geological Survey reported on this.

> There has been a striking difference in water clarity improving dramatically in Lake Erie, sometimes four to six times what it was before the arrival of zebra mussels. With this increase in water clarity, more light is able to penetrate deeper allowing for an increase in aquatic plants. Some of these macrophyte beds have not been seen for many decades due to changing conditions of the lake mostly due to pollution. The macrophyte beds that have returned are providing cover and acting as nurseries for some species of fish.[9]

Invasive plants such as phragmites and water hyacinth have been used professionally for phytoremediation efforts: they have cleansed wastewater and sewage for municipalities, corporations, and government agencies around the world. There are not only economic gains in using plants that are effective at removing heavy metals and radioactive substances, but using them is also easier, brings greater public acceptance, minimizes further spread of contaminants, and preserves topsoil. In addition, in some cases the metals accumulated in the plants can be smelted out (a source of biofuel) and recycled for further use (another source of income). According to Rufus Chaney, an Agricultural Research Service agronomist who specializes in weeds that remove heavy-metal contaminations, the cost of using plants to clean polluted soil "could be less than one-tenth the price tag for either digging up and trucking the soil to a hazardous waste landfill or making it into concrete."[10] The cost benefit of using plants is noticeable in the cleanup of lead-contaminated sites; the conventional excavation-landfill approach to such cleanup costs one hundred fifty to three hundred fifty dollars per ton, whereas phytoremediation treatment, including off-site disposal of biomass as hazardous waste, figures closer to twenty to eighty dollars per ton.

The U.S. EPA classifies some thirty thousand government sites covering more than two hundred thousand acres as containing contaminated and hazardous soils and waters. Conventional techniques for remediating toxic environments rely mainly on the transportation of massive amounts of contaminated soil to hazardous waste landfills, with an average cost of one million dollars per acre. The sites under Department of Defense jurisdiction cover twenty-six thousand acres and have cost twenty-five billion dollars to clean up. Further, the Department of Energy will use half of their two hundred thirty billion dollars to deal with the radioactive contaminated water and soil from nuclear reactor sites—they will dump it in desert wastelands. The most contaminated, hazardous, and infamous Superfund sites cost 16.5 billion dollars to clean, using current methodologies, and private industry spends an estimated twenty-four billion dollars on some twenty-four thousand sites—with mediocre oversite of each job. Even with this grand expenditure, some places are deemed too expensive to rehabilitate with conventional remediation methods, and therefore they remain toxic sites,[11] though there are prolific plants in many of these places already, providing a cleaning service and potentially billions of dollars in savings to the government (a.k.a. taxpayers). In chapter 7, we will further explore the plants that clean our toxic environments.

The ideal chemist of the future will not be satisfied with humdrum day-to-day analysis, but is one who dares to think and work with an independence not permissible heretofore, unfolding before our eyes a veritable mystic maze of new and useful products from material almost or quite beneath our feet and now considered of little or no value.

GEORGE WASHINGTON CARVER

And the future is now . . .

You never change things by fighting the existing reality. To change something, build a new model that makes the existing model obsolete.

BUCKMINSTER FULLER

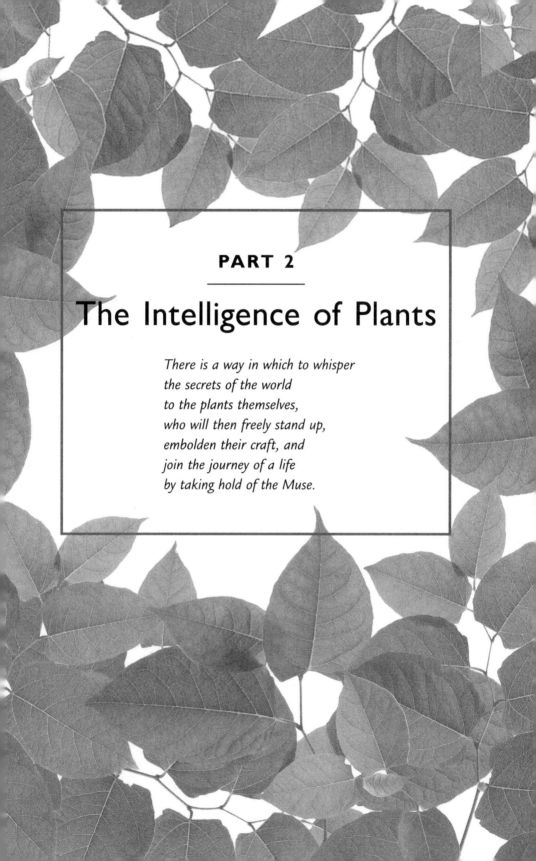

PART 2

The Intelligence of Plants

*There is a way in which to whisper
the secrets of the world
to the plants themselves,
who will then freely stand up,
embolden their craft, and
join the journey of a life
by taking hold of the Muse.*

THE INTERCONNECTION BETWEEN PLANTS and people runs from a primal source, though most of the time, we humans go about our day forgetting the fundamental need for plants in our modern lives. Even before we came into existence as humans, plants were the life force on this planet and were formed to absorb and transform vital radiation from the sun. The plants eventually nourished and sustained hungry foragers who came along, with their green nectar of solar essence collected in photosynthesizing leaves. As they enter the bodies of animals, plants assume a two- or four-legged form as their mode of transportation, and synthesis of the two kingdoms occurs.

> *Plants are the placenta of animal life. . . . Every carbon atom in our bodies has at one time passed through the chloroplast membrane of a plant.*
>
> DALE PENDELL, *PHARMAKO/POEIA*

Plants have captivated us and lured us to move them for eons. Our dependence on them extends so deep into the primordial soup of life that Pendell muses, "only plants had consciousness. Animals got it from them."[1] Plants seem to have a plan to spread throughout the land wherever they can and breathe life into the soil and bear fruits (with seeds) so that these may be consumed and may spread farther. Plants provide for all the species that pollinate, eat, or use them in any way—both for their own benefit and for a kind of purposeful design. To plants, then, we humans must be no different from the birds and the bees. We're just another vector for distribution or another symbiosis of species.

> *I dream with the plants. . . . They speak in tones of feeling and express themselves in distinct forms. They are the interdependent, breathing follicles on the skin of the earth, receiving and vitalizing light so that it can become energy, storing this energy deep within themselves:* transmuting, transforming, purifying, and balancing. *Day and night, they perform myriad effortless tasks that are as simple as*

88

breathing but that have complex chemistry. Doing what only plants can do: *they help to stabilize habitats and create integrity within the landscape by enriching the soil with their growth and by not allowing further soil disturbance. They transform industrial runoff, pharmaceutical pollution, and other toxic pollutants. The city-dwelling weeds are breaking through the concrete, allowing the earth to breathe; they are climbing the cement walls in order to cool the city jungle and revitalize the air. They are the reckoning, cleanup crew that serves the greater organism of the planet, clearing the way for a new evolution of this living being we call Earth.* No environments ever stay the same. *Only we humans may not perceive the changes, because we tend to focus on ourselves in a definable timetable: a lifetime and how to survive it.*

So why should this next evolution of changing environments be so different? Is it because we humans are involved? Are we so different from a great ice sheet moving across the land, and then retreating, altering everything it touches? Or, as some surmise, are we a plague? Or are we simply another expression of Kali, the Hindu goddess of destruction?

Death in order to create new life . . . with Nature leading the proper course of reclamation.

Rampantly endemic plants seem to touch many individuals in vulnerable places, and people's response to this exposure is sometimes dramatic. Plants have entered backyards at night—a scary premise to many people—so we build fences to keep out those plants that are wild and rambunctious, and we make and apply poisons to tame them—but in the end, Mother Nature endures.

These ubiquitous plants represent true wildness in our lives, and there is no controlling them. They move in drastic ways, marching on in disproportionate numbers as if chanting and calling for attention. People, however, have not stopped shouting back, and the plants' message is left unheard. The wildness they represent as defined by us brings up fear for some people, and some human reactions to uncivilized nature are often extreme. The strong response of some people regarding these untamed expressions of nature reveals a potent relationship with what is natural or wild.

Plants that take over the land mirror human behavior: we have taken over the land, covering it with concrete and asphalt, with hardpan parking lots that spread across millions of square miles of the earth. Throughout these human-constructed environments, it is hard to find anything resembling native soil, and diverse natural habitats are paved over.

> But some are beginning to break through this crumbling illusion. . . .

Dandelion, tree-of-heaven, plantain, and other pioneering plants find the cracks and adapt well to the harsh environs of a concrete jungle. They seem to be following us, and they shadow our footsteps.

> The weed that will not go away and that you continually notice in irritation, the plant that regularly trips you as you walk through the field. Such plants are often some of the most powerful medicines you will find. They stir something in your unconscious, breaking through your habituated not noticing, and intrude on you until you begin to take a real look at them.
>
> STEPHEN HARROD BUHNER,
> *LOST LANGUAGE OF PLANTS*

At times, however, these invading plants seem even to precede the charge: often, potent medicinal plants colonize territory before the invasion of the very pathogens and toxins they cure. It is as if the plants anticipate their duty as instrumental healers of the land and its inhabitants and arrive in advance of our need for them. Invasive plants often have a head start on the intruding pathogens and toxins that sneak past our immune defenses and settle into our bodies as chronic disease. Therefore, it is of little service to ourselves if we continue to battle these plant allies. Instead of enjoying the benefits of the potent medicine offered by these plants, people fight them and continue to fall ill from organisms that grow resistant to our overuse of antilife weaponry—just as plants grow resistant to our herbicidal onslaught.

The vigor of these plants—even against being violently removed—reveals much about the potency and actions of the medicine they provide for both the land and people.

Oracles

Though the gods have the power of speech
more often they choose a flower or plant;
elder leaves pressed on a blotter,
or spring buds emerging from a winter stem.
These messages they send—so ordinary we usually miss
*   them . . .*
The subtlest oracles are always the most obvious—
seeing what's clearly in front of us the most difficult:
a butterfly hatching from a ruptured dream,
or a splintered tree rooting in the soil where it fell . . .
The god's whispers are never commands,
more like the place a steep trail has collapsed,
and sunlight offers the understory a second chance.

DALE PENDELL,
LIVING WITH BARBARIANS

7

The Deep Ecology
of Invasives

The essence of deep ecology is to ask deeper questions.
ARNE NAESS, FOUNDER OF THE DEEP ECOLOGY
PHILOSOPHICAL MOVEMENT

DEEP ECOLOGY FOCUSES ON studying the endless interconnections between all species (including humans) and their environments. This study involves the physical changes, adaptations, and links within ecosystems as well as, the mental, emotional, and spiritual dependence between humans and the infinite manifestations of the living mind of Earth. When we examine a single species with this mode of thinking *and feeling,* an expansive ripple of concentric rings passes through all forms of life in the area, touching microbes, fungi, plants, animals, and landscapes. In turn, the other species of life within range return chemical vibrations that are felt by this single life-form in many invisible ways.

In order to clear our minds and calm our emotions, our *spirit-heart-mind* feeds on the vibrational essence of the wild. We depend on it as a source of sustenance—our soul food. We absorb a wild essence from the feeling we get from old-growth forests, majestic mountains, or the ocean breeze. In city environments, our life force is rejuvenated and replenished with the natural energies provided by parks that serve as a kind of oasis in the desert. Or as James Lovelock conveys, "They seem to serve as reference points of health against which to contrast the illness of the present urban and rural scene."[1] We are at once cleansed, refreshed by the vibrant wilds, and sent into a place of deep remembrance of an ancient, native self. This is a medicine, and when we absorb it, it reverberates throughout our being and mends a severance at our core. In turn, we can provide this reverberation to others as a potent life force—as evidenced by the times when we see in the eyes and feel from the hearts of great healers and teachers. When our spirit-heart-mind is connected to our gardens, plants respond to this energy and flourish. (Plants also like Vivaldi and Beethoven.) Humans are always "doing" to the environment as well as receiving from it—and the result is a subtle, deep impact on all levels of life.

> *Life is a web. What we do to the web, we do to ourselves.*
> CHIEF HIAWATHA, FOUNDER OF THE
> IROQUOIS CONFEDERACY

All unfolding ecological systems have complex interactions among the various species of life that arrange complete systems within the soil and above ground. In the rhizosphere (soil surrounding plant roots), microbial and mycorrhizal life set the foundation for plant life by helping to provide nutrients, water, and other chemicals that plants take in and utilize for growth. Within healthy forest soil, we find hundreds, if not thousands, of different life-forms in a square yard of earth. In this small world, one species provides for another, which in turn provides for another, and so on, through cycles of life and death. In addition to the plant-supporting life underground, a vast array of animals and insects helps reproduce these plants and spread their offspring far and wide. Pollinators, consumers, and carriers all perform vital roles.

Because of this intricate interdependence, if one life-form is removed from the environment, very often, other life-forms suffer as a result. For example, removal of the mycelial (fungal) network leads to fewer nutrients and reduced water uptake for the plants. Therefore, the entire ecosystem as a whole is weakened. If any plant species disappears, a new niche opens, requiring adjustments in the environment. This occurrence sets free highly adaptive propagules (plant offspring) to fill in the gaps.

Yet with so many life-forms providing benefit to others, how can we separate the threads of life? Most scientific methods focus on an individual part of nature. Without the freedom to observe broadly the interconnections among different life-forms, how are we to distinguish fully the consequences of adding species to and subtracting species from an ecosystem? Though relatively new to a landscape, opportunistic plants have already developed within a complex community of species, cooperating with the ecosystems they inhabit to address any available needs and niches.

> *Plants and soil exist in a tightly linked partnership. Just as plant growth cannot be understood without reference to soil, neither can soil structure or the processes occurring within soil be understood without reference to plants.*
>
> DAVID PERRY, *FOREST ECOSYSTEMS*

A dichotomy has impregnated predominant ideology with regard to the natural world around us. Over the past few hundred years, the scientific method has broken down into parts every bit of nature and has viewed each as a disparate piece. This has led to perpetual conflict between those in the mechanical, scientific, and technological world and those who live with a more holistic, intuitive, integrated view of the natural world. Humanity has learned much by dissecting the world into pieces and trying to find the basis of life in the smallest speck, but this has revealed an incomplete understanding that is riddled with misconstrued impressions. Many individuals have integrated both sides of the conflict and have come away with a deeper, more comprehensive understanding of life. In order to have a richer, fuller picture still, we must further blend these two pools of understanding. It is much like the common example of understanding a piece of artwork: a scientist can look at a beautiful painting and analyze the different chemicals in the paint and how they interact and blend. In the end, though, this approach will never lead to the true meaning of the picture. To grasp the whole meaning, we must step back and observe completely the lines and colors and see the beauty contained within. This is what Deep Ecology is about: taking in the full tapestry of nature, with all its interrelatedness, to absorb the gestalt present there. As we have seen throughout history, the people who have integrated parts and expanded their vision in such a way have made monumental changes to the way common science observes phenomena.

> *Look deep, deep, deep into nature, and then you will understand everything.*
>
> ALBERT EINSTEIN

ECOLOGICAL HEALTH MIRRORS HUMAN HEALTH

As James Lovelock and Lynn Margulis's Nobel prize–winning Gaia Theory conveys, Earth is a living organism of interdependent biosystems that feed one another in order to sustain the whole of life on the planet. Gaia has temperature regulation, circulation, respiration, and metabolism just as is found in many of her organisms. The corporeal system of Gaia comprises different landscapes and ecosystems that function as ecological organs—oceans, rivers, forests, mountains, and wetlands. With the aid of the sun, the oceans regulate temperature, creating clouds and rain to shade, nourish, and cool the earth. Springs, streams, and rivers form a circulation system that carries the lifeblood throughout the land. Mountains erode and volcanoes erupt over time and slowly release various chemical vapors, generating a unique atmosphere of "extraordinary and unstable mixture of gases,"[2] which leads to a perfect balance for life to take place. The earth does breathe in a way, with the inhalation and exhalation of oxygen and carbon dioxide throughout the forests (lungs) of the planet. Wetlands and bogs act as the earth's liver, purifying and cleaning toxins in the water and controlling methane emissions from the fermentation of organic plant material. Gaia has been fully self-regulated by all life on earth for hundreds of millions of years, and our unprecedented time in the planet's history is requiring extreme adjustments of ecosystems. The earth must develop ways of purifying the toxic overload created by the modern industrial age.

As we are discovering, our ecological systems are struggling to cope with the stressors we put on them. The forests are becoming weaker and more susceptible to disease as a result of pollution and improper forestry methods. Modern agricultural practices have cleared the precious topsoil, devalued the nutrients from the soil and our food, and intoxicated the land and water. In the War on Pests, companies are producing stronger pesticides and herbicides in greater quantities in an attemp to prevent a complete takeover by virulent invaders, yet these plants and pests are growing ever resistant to these poisons.

The degradation of the health of the whole as an unintended consequence of futile attacks on invaders is not an isolated phenomenon, but instead it is a pattern that repeats on the macrocosmic as well as microcosmic scales. The deteriorating health of our forests is analogous to the current weakening of the human immune system. Widespread chronic disease, antibiotic-resistant bacteria, and emerging, endemic diseases result from modern medical practices just as Gaia's infected biosystems are caused by modern industrial and agricultural practices. Invasive epidemics are based on the widespread use of toxins and poisons that infect all biosystems, great and small, and these destructive influences trickle down into cellular life, mortally impregnating all Earth's species with deformative and abortive destiny.

THE PLAGUE OF POLLUTION

It is impossible to escape the pollution of the day—virtually everything we touch and make ties in to the source of its creation. We have filled volumes of books describing the impacts of the various contributors of pollution and their toxic emissions into our surroundings. The air, soil, and water, which all inhabitants of this planet use, are alarmingly toxic. Even the most pristine, remote places are infected with heavy metals, toxic chemicals, acid rain, and pharmaceuticals. Essentially every product that goes through a manufacturing process ends up creating by-products that are released (gases, sewage, wastewater, radioactivity) into the environment, both near and far.

Lately, there have been many studies detailing how chemicals end up in our water supplies. In one U.S. Geological Survey study, one hundred thirty common chemicals remain in drinking water supplies across the United States, even after standard treatment processes, and this presence affects hundreds of millions of people through the kitchen-sink tap and the bathtub. The chemicals include pesticides, gasoline hydrocarbons, household-use products and solvents, and a vast array of pharmaceuticals—and most of these are unregulated by the government, which does not require monitoring or removal of them from public drinking water supplies. These chemicals have also been found in many brands of bottled water: the Environmental Working Group released a study by the University of Iowa Hygienic Laboratory that detected thirty-eight toxic pollutants in the ten leading brands of bottled water, causing the organization to report that "bottled water was chemically indistinguishable from tap water."[3] In addition to the chemicals mentioned here, also found in the bottled sources were fluoride, by-products of chlorine-based disinfection, bacteria, caffeine, arsenic and other heavy metals, and radioactive isotopes. Additional research performed by the University of Missouri found that bottled water can increase the proliferation of cancer cells because of its estrogen-mimicking effects.[4] All this seepage is creating a toxic mix of synergizing molecules, which form complex chains of reactions and lead to a dangerous brew that we drink straight from the public water utility and from the multibillion-dollar bottled water industry.

And if it is in our water, it is everywhere, for water is the basis of life as we know it.

People are directly ingesting these chemicals through our food supplies; the chemicals leach into the plants that are swallowed by the animals we consume. High levels of heavy-metal contamination result from fertilizer and pesticide use, and increased bacteria and foodborne pathogens result from antibiotic overuse. Indeed, outbreaks of these pathogens in our food supplies are reported all the time. Along with conventional produce, many products on the shelves contain GMO corn, soy, and wheat that are inundated with increasing pesticide residues. (Further, these products do not have to be labeled GMO, according to the decree of the Food and Drug Administration.) Studies funded by the Environmental Protection Agency have found in U.S. fish supplies pharmaceuticals that treat high cholesterol, allergies, high blood pressure, bipolar disorder, and depression; household chemicals; and toxic heavy metals (most notably mercury).[5] These chemicals eventually find their way into our bloodstream, where they slowly deteriorate basic bodily functions and replace with toxic ones the vital elements that are essential for life.

Sources of Pollution

Natural Sources
- Earth, rocks released through weathering and erosion
- Volcanic gases

Industrial Sources
- Chemical manufacturing
- Metal manufacturing
- Paint and dye
- Petroleum industry
- Pharmaceutical manufacturing
- Leather manufacturing
- Textile, paper, plastic, rubber manufacturing

Electricity Production
- Coal
- Nuclear
- Wood

Hospital and Medical Sources
- Hazardous waste removal and incinerators
- Pharmaceuticals
- X-rays, radiation, chemotherapy
- Sterilization and cleansing agents
- Wastewater

Domestic Wastewater
- Cleansing agents
- Pharmaceuticals
- Cosmetics
- Human waste

Agricultural Sources
- Pesticides, herbicides, chemical fertilizers
- Antibiotics, hormones
- Animal waste

Vehicles and Heavy Equipment
- Emissions from cars, trucks, trains, boats, airplanes, tractors, and other heavy equipment
- Oil, gasoline, radiator fluid, and other fluids
- Road salt
- Batteries and electrical components with toxic, reactive metals

Mining
- All metals are mined and brought to the surface in concentrated forms
- Runoff from rain and the cleaning process of the metals flows downstream

Solid Waste Disposal and Landfills
- Computers, lightbulbs, and whatever else we throw away
- Concentrated waste of various sorts, including toxic cleanup from other sites

Nuclear Industry
- Radioactive waste and emission, wastewater

Military Industry
- Ammunitions, explosives, gases, nuclear waste

Atmospheric Pollution
- Acid rain
- Smog

Electromagnetic Sources

- Electricity, microwaves, radiowaves, cell phones, and so forth

Note: I cannot here address the subject of a plant's ability to effect electromagnetic pollution, but such pollution is another silent poison that impacts on elemental microscopic levels. I have experienced some success in using yarrow flower essence (and Flower Essence Society Yarrow Special Formula) to assist in fortifying the protective sheath—an electromagnetic immunomodulator—especially for those who are overly sensitized to the pressures of techno-societal spheres of influence, as found in the electromagnetic and emotional fields throughout our environment.

It is overwhelming to try to comprehend and codify the totality of our toxic surroundings. It is hard to imagine that we are still alive at all with the complete dissemination of poisonous substances touching all life on Earth, inside and out.

> *We are fortunate to have the wisdom of nature . . . to counterbalance the foolishness of humans.*

MEDICINE FOR THE EARTH

Medicine

As dreams are the healing songs
from the wilderness
of our unconscious—
So wild animals, wild plants, wild landscapes
are the healing dreams
from the deep singing mind
of the earth.

DALE PENDELL, *LIVING WITH BARBARIANS*

Phytoremediation: From Sewage to Flowers

The increased heavy metals and toxins present in our surroundings have put at risk the world's biosphere (all of life). Fortunately, the earth has an innate ability to work with plant life to clean contaminants from polluted ecosystems. Phytoremediation was founded in the 1990s, and, although a fairly new science, numerous studies from around the world confirm the ability of the plant world to clean toxins and heavy metals from the environment. This ancient and unique capability of plants has been discovered and utilized in the practice of bioremediation, which processes the inefficient by-products of modern technology that have contaminated the land, air, and water in the first place.

Although many minerals are essential to plant growth, all metals are toxic at high levels. Plants normally take up trace amounts of metals into their root systems, but excess levels create stress many plants cannot tolerate. In a way similar to that outlined in human physiology, heavy metals in the soil and plants cause oxidative stress through the formation of free radicals, and this slowly destroys the basic functioning of cells. The metals can also leach out or replace essential minerals within the cell and disrupt the basic workings of enzymes and pigments. However, many plants have been found to create their own metal-chelating agents—*phytochelators*—that help the plants process and eliminate these toxins, just as certain compounds are used in chelation therapy for metal poisoning in humans. Some of the measurable compounds known to help detoxify and moderate the metals in plants are metallothioneins, phytochelatins, phytins, organic acids, and amino acids with, often, a sulfur component. Other animals, fungi, and microscopic life-forms use similar complexes and make use of these plant compounds for toxic accumulations of their own.

Complex interactions take place within the soil among plant roots, microbes, and metals that facilitate the process of cleaning heavy-metal contamination. Within the rhizosphere, surrounding microbes interact with compounds that are secreted from plant roots, thereby assisting the elimination of metals from the soil. Chemical plant compounds work as antioxidants for the soil: they scavenge heavy-metal free radicals, helping to bind and neutralize the toxins. Some of the known effective bioremediation methods use microbes and fungi to break down organic compounds and accumulate heavy metals from the environment. According to Paul Schwab, phytoremediation expert and director of natural resources and environmental science at the Purdue University, "[T]he plants accelerate the microbes' action in the soil, [and] they stimulate microbes to degrade contaminants by getting more oxygen into the soil and by supplying nutrients through their roots."[6] In many cases, plants appear to be quite effective at remediation efforts and are an impetus for obvious as well as unseen benefits.

Phytoremediating plants are able to gather large quantities of these contaminations in their root systems. Some store it there for a while, and others send the metals into the plant body (leaves, stems, and fruit) to be captured for harvest. Many people have noted that high-level metal-accumulating plants (hyperaccumulators) are markers of high-metal and chemical soils, because the hyperaccumulators tend to dominate the land where certain metals accumulate. These plants were observed by German scientists in 1855 while they studied *Viola calaminaria* in the zinc-rich soils of the Aachen area, bordering modern-day Belgium. Others, too, have used plants as bioindicators to locate metal deposits in the soil in order to mine the area. The desert trumpet can indicate the presence of gold, and alpine pennycress often indicates the presence of zinc and nickel. In some phytoremediation efforts, the metals can be smelted out and recycled for further use, or the plants can render the once-toxic element inactive and safely release it back into the environment.

Using plants to clean the environment requires less money, energy, and labor, and it can benefit a community by generating a renewable resource and income. During the phytoremediation process, a biomass fuel source can be made from harvesting the plant bodies, and the metals contaminating the soil can be recycled by the remediating plants and made use of again. Employing plants in this way does not require any expensive equipment or specialized personnel. There is greater public acceptance of this practice, because it preserves precious topsoil without hauling, thereby decreasing chances of further contamination, and it creates parklike settings instead of creating a site that is inhospitable and cordoned off. Using plants for remediating purposes also reduces landfill space up to 95 percent by eliminating the need for dump truck loads of toxic soil. In areas of mining that have been left with severe surface damage and heavy-metal accumulations unsuitable for most plant growth, many tolerant invasive pioneers help revegetate the contaminated area and build soil with fast-growing organic matter. The high biomass plants such as phragmites and Japanese knotweed can accumulate heavy metals and transform topsoil over the course of time, building up soil levels so that other plants may then take root.

The greatest complaint of phytoremediation work is the time it takes to clean the hazardous sites. The process usually takes between eighteen months and five years, with other conservative estimates closer to thirteen to sixteen years for remediating a typical site. Nature works on a different timetable, and we must remember that we have deposited these toxins in extreme abundance over the course of fifty to one hundred years. This gradual return of elements to the environment, transmuted by the plants into less reactive forms, leaves minute concentrations of heavy metals and toxins in the soil. Phytoremediation skeptics reference this incomplete removal as another disadvantage of plant remediation. Yet phytoremediation, like all bioremediation efforts, does not completely remove 100 percent of the contaminants because reaction rates decrease over time as concentrations of the contaminant decrease. Even moving dump truck loads of toxic pollutants to areas deemed appropriate for long term "storage" can spread the toxins en route, creating the same problem in a different place. To remediate the trace contaminants residing in the soil, some people are combining phytoremediation with other methods. High-biodiversity environments can prove effective; combining all levels of soil life through microbial, fungal, plant, and tree forms that expand the sequestering of toxins within the soil, increase the absorption of light for photosynthesis, and assist the release and collection of contaminants through foliage.

Part of what we observe in the invasion of opportunistic plants is that they are the *bioindicators* of modern-day, wide-scale, industrialized pollutions. It was not until the twentieth century—with the creation of technologies that extracted metals from the earth in high concentrations never found in nature—that we saw greater emergence of these plants. Both the contaminants and the plants have entered the landscapes and waterways around us. The plants spread farther, mirroring and paralleling large deposits of heavy metals and radioactive elements that leach into our environments: lead in the 1920s; selenium in the 1930s; nickel in the 1940s; cobalt and copper in the 1960s; cadmium and manganese since the 1970s; and uranium, cesium-137, and strontium-90 in the 1980s. Fortunately for us, the planet has adapted and learned to process these corruptions through the inexplicable genius of nature and the unique healing abilities of its plant stewards.

> *By now, we shouldn't be surprised. . . . Monsanto, DuPont, Chevron, and other industrial chemical and petroleum companies have already had their hands on this one, manipulating nature again, developing phytoremediation techniques by gene-tweaking plants so that they become hyperaccumulators and extract from soil the heavy metals that these companies put in the soil in the first place.*

Types of Phytoremediation

Phytoextraction is the means of taking heavy metals from the soil into the aboveground parts of the plant via the plant's roots. The plant is then harvested as biomass and transported to incinerators or waste sites. For areas of extreme toxicity, hyperaccumulators are the plants sought: they grow quickly and store high levels of metals in their leaves and stems. In a study appearing in *Nature,* brake fern (*Pteris vittata*) was highlighted by researchers at the University of Florida as an arsenic-lover, thriving in laced environments and accumulating "with staggering efficiency"[7] the carcinogenic heavy metal. This has led to further utilization of this plant to clean arsenic from polluted land. Widespread release of arsenic as a result of mining, milling, combustion, wood preservation, and pesticide and herbicide use (especially for golf courses and lawns) has created tens of thousands of known arsenic-contaminated sites around the world.

With this process, some researchers and companies *phytomine* con-taminated land and reuse the toxic minerals by transforming them into valuable, high-grade ore. USDA agronomist Rufus Chaney and colleagues at the University of Maryland patented such methods of biomining in 1998. Chaney explains the process:

> The crops would be grown as hay. The plants would be cut and baled after they'd taken in enough minerals. Then they'd be burned and the

ash sold as ore. Ashes of alpine pennycress grown on a high-zinc soil in Pennsylvania yielded 30 to 40 percent zinc—which is as high as high-grade ore. Electricity generated by the burning could partially offset biomining costs.[8]

He goes on to describe the ideal plants to do this: "They'd have all the characteristics of a hay crop: They should be tall, high yielding, fast growing, easy to harvest, and deep rooted. And they should hold onto their mineral-rich leaves so they can be harvested along with the plant stems."[9]

Rhizofiltration is the process of collecting contaminants in water-saturated roots through wastewater systems and polluted wetlands and rivers. Most of the plants that grow in these areas filter and process huge volumes of water. During the water intake, heavy metals and contaminants present in the substrate are brought into the plant body and sequestered, metabolized, or vaporized through the foliage. Plants that take up large amounts of water are ideal candidates for rhizofiltration, for they capture and immobilize large amounts of toxins through the water in wetlands and rivers. Plants such as reed, cattail, Japanese knotweed, tamarisk, and purple loosestrife all have these desired capabilities and have been used for this purpose.

Phytovolatilization utilizes plants to extract from the soil dangerous volatile metals (such as mercury and selenium), transform them into less harmful forms, and release them in gaseous form through the foliage and into the atmosphere. English ivy and tamarisk have been found to be able to clean the soil and air, releasing volatile elements through their leaves. Some plants then employ *phytostabilization*—stabilizing pollutants in the soil so that further contamination of surrounding areas and groundwater is halted. Hybridized poplars and willows are commonly used commercially for this purpose.

The most informative field trips I take to find invasive species are along highways and roads. In every region of the country, we can discover these plants growing in abundance throughout the gravelly medians and along the roadsides. Some spread out into the surrounding wetlands and culverts and others climb up the forest edges. In the northern United States, Japanese knotweed, purple loosestrife, and common reed are usual roadside companions. Often, some bittersweet clears the canopy on trees that grow on the edge of the road. Farther south, a traveler enters the territory of mile-a-minute that climbs the concrete walls and kudzu that spreads out along the roadways and topples the forests. In the western states, carpets of thistle, leafy spurge, and knapweed travel for miles, covering the vast rangelands, and Scotch broom holds back the eroding roadsides, creating a mesmerizing cloud of yellow when in bloom. The abundance of these plants along roadways leads to an understanding of the significance of allowing these plants to grow where they do.

Areas around highways must be some of the most polluted . . .

When I contacted Paul Schwab to garner information on phytoremediation work with invasives, he informed me of the term I had been seeking: *natural attenuation.* This is a type of phytoremediation in which the ecosystem's innate toxin removal faculties take over without human interference. The Environmental Protection Agency defines natural attenuation as

> a variety of physical, chemical, or biological processes that, under favorable conditions, act without human intervention to reduce the mass, toxicity, mobility, volume, or concentration of contaminants in soil or groundwater. These in situ processes include biodegradation; dispersion; dilution; sorption; volatilization; radioactive decay; and chemical or biological stabilization, transformation, or destruction of contaminants.[10]

In fact, the discovery of plants growing in contaminated environs is the common means scientists use to determine the potential for remediation.

Unfortunately, this principle has been abused by many companies as they reference the natural cleansing process as an excuse to sidestep serious remediation efforts. With human distortion of cleanup efforts set aside, the work of Mother Nature has excelled: Nature has offered the green hazmat team as a response remedy to our perpetual so-called advancements in industry and technology. As human attentions focus on plant remediation in order to find cures for techno-industrial ills, it is becoming clear that plants are already in place and doing what the researchers are exhaustingly trying to re-create in the labs.

Remember: The plants grow where they do for a reason.

Invasive Plants as an Ecological Remedy

Evidence is mounting that the vigorously growing blends of native and non-native plants that "invade" damaged land are yet another example of nature's wisdom and resourcefulness. Nature creatively mingles both native and exotic without prejudice, using all resources available to throw a green Band-Aid over ravaged landscapes.

TOBY HEMENWAY, *GAIA'S GARDEN*

Clearly, there is reason and intention behind the settlement of opportunistic plants in damaged and disturbed ecosystems. As bioindicators and hyperaccumulators, plants serve as translators and healers between humans and the invisible forces of nature. At times the rambunctious invasive species rampantly parade with bells and whistles, trying to get our attention and warning us of our toxic follies. It so happens that at the same time, other less adaptive native species, with their niche disrupted, are disappearing because they are unable to filter pollutants as the invading plants can. The opportunistic healing plants are filling in the gaps and softening the edges and wounds of roadways, clear-cuts, and developments that humans create. The sharp, painful contrasts that are created when ecosystems are penetrated and disturbed must be alleviated by plants such as Oriental bittersweet, kudzu, and mile-a-minute. These vines create a netlike barrier to prevent further intrusion of humans or machines. Some of these plants are extremely protective, as Amy Steward describes in *Wicked Plants*: "At the Fort Pickett military base in Virginia, kudzu overwhelmed two hundred acres of training land (and) *even M1 Abrams battle tanks couldn't penetrate the rampant growth*" (my italics).[11] The thorny invasives also keep disturbing elements at bay: wild rose, blackberry, and barberry protect the beaten land and forests, controlling erosion runoff, and allow the place to rest and rejuvenate with time. Others, such as knapweed, wild mustard, and star thistle move in and dig deep to protect, enliven, and rehabilitate the baked, barren soils of long-abandoned farmland that has subsequently been trampled by grazing livestock. Cornflower and bindweed can appear in the fields, alerting the farmer of improper agricultural practices and exhausted soils. Other plant invaders such as Scotch broom and Russian olive enrich soils and provide valuable resources to desecrated, parched environments and have even been employed by humans for just such a purpose. Plants also gravitate to modern-day disturbances that remain invisible to the human eye; they naturally attune to areas harboring excess contaminants and lessen the toxicity. Invasive plants are the prescribed remedy of the industrial age, adapting to and mitigating the symptoms and side effects of modern progress.

Another field observation took place at my friend's home in Northampton, Massachusetts. Sigrid and Winston moved into a house that was heavily contaminated with lead paint, and the backyard was used as a garbage bin by the previous owners going back generations. My friends have had to renovate nearly the entire house to make it child-safe—and this has also meant a long, arduous, and dirty mess to clean up. While standing in the yard and admiring the knotweed growing as a hedge, I began to look around and was excited to see other plants. Their yard was an invasive plant playground. In addition to the extensive knotweed, there were garlic mustard, plantain, pokeweed, mustard, blackberry, burdock, yellow dock, and ivy. A couple of years ago Winston cleared out Oriental bittersweet to stop it from overtaking four trees along the yard, and according to him, "They were all just growing here; we didn't do a thing." Yet in the surrounding yards and wild areas there wasn't the same distinction of species. This kind of situation might have thrown some purists into a frenzy, yet as I stood in their yard, a thrill tickled up my spine, and I felt honored to know the importance of these plants' presence.

To me, it was reassuring to know they were there.

The fear of greater spread and the natives-only approach have limited some remediation studies that focus on using invasive plants in the United States, yet surprisingly, from their regions of origin, a large amount of research examines these plants. Many individual plants have been investigated with remediation purposes in mind, and all together, these oppor-

Sigrid and Winston's invasive plant playground

tunistic species show abilities to clean a vast array of toxic chemicals and pollutants from the soil, water, and air.

In one study of 5 rivers and 236 sites (each site being 500 meters long), with Japanese knotweed growing along the banks, the plant was found to grow with the highest frequency and abundance in the areas of land with wastewater runoff, instead of in so-called natural, seminatural, or grazed land, and this plant has been found in another study to accumulate zinc, lead, and copper in its plant body. Purple loosestrife has been found to soak up excess nitrogen and phosphorus buildup from fertilizer runoff into waterways. Kudzu cleans petroleum and chromium contamination from spills and scrap yards, and English ivy has been found to purify the air of benzene and toluene in polluted cities and indoor environments. In numerous studies, tamarisk (salt cedar) exhibits an array of cleaning capacities; it removes cadmium, lead, copper, arsenic, and sodium. The bountiful dandelion and blackberry accumulate zinc and copper, and dandelion is also able to clean manganese, lead, and cadmium from the environment. Knapweed absorbs radiocesium and nickel that has silently accumulated in western rangelands.

And the granddaddy of them all: the electronic file of scientific studies regarding common reed (*Phragmites*) was so large that it had to be sent to me from Schwab separately from the file for other plants. Phragmites has been used for treatment of mine wastes; explosives; and agricultural, industrial, and municipal wastewaters throughout the world. Under scientific scrutiny in hundreds of studies, common reed displays the ability to effectively clean sewage wastewater and remove fifteen heavy metals (Zn, Ph, P, Pb, Cd, Cr, Ni, Mn, Cu, Zn, Se, Na, Th, U, Cs-137) and at least eleven common toxic pollutants (herbicides, petroleum, TNT, DDT, PCBs, xenobiotics, chloroben-zene, phenols, sulphide, acid orange 7, polycyclic aromatic hydrocarbons). It is no wonder that we find this plant growing in abundance along roadsides, drainage ditches, floodplains, and wetlands, since this plant naturally attunes to and attenuates these polluted water catchments. Reed serves as the chief water filter and detoxifier throughout the earth's cleansing wetland systems. Some of the following brief synopses of studies exploring invasive plants' ability to clean contaminants were located for me by Paul Schwab.

Scientific Studies

Phytoremediation of Explosives in Toxic Wastes

Selected emergent plants *Phragmites australis, Juncus glaucus, Carex gracillis,* and *Typha latifolia* were successfully used for degradation of TNT (2,4,6-trinitrotoluene) under in vitro conditions. The plants took up and transformed more than 90% of TNT from the medium within ten days of cultivation. The most efficient species was *P. australis,* which took up 98% of TNT within ten days. . . . The results were verified in pilot constructed wetland for cleaning explosive containing waste-wa-ters as a necessary step prior real scale-up application.

Source: I. Twardowska, H. E. Allen, M. M. Haggblom, S. Stefaniak, "Phytoremediation of Explosives in Toxic wastes," *NATO Advanced Research: Viable Methods of Soil and Water Pollution Monitoring, Protection, and Remediation* 69 (2006): 455–65.

Detoxification of Herbicides in *Phragmites australis*

Unintentional loss of herbicides into drainage ditches, shores, or other waterbod-ies may cause large problems in farmland. Therefore strategies for the phyto-remediation of agrochemicals and especially herbicides have become a topic of great interest in many agricultural areas. However, in order to establish effective biological pollution control, information on the detoxification capacity of riparian plants and aquatic macrophytes (e.g., *Phragmites australis*) is important to build up effective buffer strips. We determined the detoxification capacity of *Phragmites australis* roots and leaves for the conjugation of agrochemicals to glutathione by assaying the model substrate CDNB as well as the herbicides fenoxaprop-P, propachlor, pethoxamid, and terbuthylazine. . . . In summary, *Phragmites australis* seems to be efficient in herbicide detoxification and a good candidate for phytoremediation of effluents from agricultural sites.

Source: P. Schroder, H. Maier, and R. Debus, "Detoxification of Herbicides in *Phragmites australis*," Zeitschrift Fur Naturforschung section c-A Journal of Biosciences 60, no. 3–4 (2005): 317–24.

Lead and Cadmium Accumulation and Phyto-excretion by Salt Cedar (*Tamarix smyrnensis*)

The accumulation and excretion of lead (Pb) and cadmium (Cd) by salt cedar (*Tamarix smyrnensis*) were investigated in this study. . . . The excretion of Pb and Cd by salt glands was observed and quantified. *T. smyrnensis* excreted a significant amount of metals on the leaf surface. This characteristic of salt cedar plants can be viewed as a novel phytoremediation process for the remediation of sites contaminated with heavy metals that we have termed "phyto-excretion."

Source: J. Kadukova, E. Manousaki, N. Kalogerakis, "Pb and Cd Accumulation and Phyto-excretion by Salt Cedar (*Tamarix smyrnensis*)," International Journal of Phytoremediation 10, no. 1 (2008): 31–46.

Efficacy of Indoor Plants for the Removal of Single and Mixed Volatile Organic Pollutants and Physiological Effects of the Volatiles on the Plants

Foliage plants of English ivy (*Hedera helix L.*), peace lily, nephthytis, and grape ivy were evaluated for their ability to remove two indoor volatile organic air pollutants, benzene and toluene. . . . *H. helix* was substantially more effective in the removal of either benzene or toluene than the other species, with the removal of toluene more than double that of benzene.

Source: M. H. Yoo, Y. J. Kwon, C. K. Son, K., S. J. Kays, Journal of the American Society for Horticultural Science 131, no. 4 (July, 2006): 452–58.

INVASIVE PLANTS AND THEIR KNOWN BENEFITS TO THE ENVIRONMENT

PLANT	EROSION CONTROL	REMOVES TOXIN	REMOVES HEAVY METALS	CLEANS WASTEWATER	CLEANS AIRBORNE POLLUTANTS	PROVIDES WILDLIFE FOOD	PROVIDES BEE NECTARY	ENRICHES SOIL	PROTECTS EXHAUSTED & TRAUMATIZED LAND
artemesias	yes	petroleum, explosives (HMX)	Pd, Zn, Cu				yes		
barberry						yes	yes		yes
bindweed	yes	petroleum	Cd, Cr, Cu, As, Sb, Tl, Zn			yes		yes	yes
blackberry	yes	herbicide (MSMA)	Zn, Cu			yes	yes		yes
dandelion		yes	Cu, Zn, Mn, Pb, Cd				yes	yes	
English ivy					benzene, toluene	yes			yes
garlic mustard						yes	yes	yes	
Japanese honeysuckle						yes	yes		yes
Japanese knotweed	yes	yes	Cu, Zn, Pb	yes				yes	yes
knapweeds		radiocesium	Ni				yes		yes
kudzu		petroleum	Cr			yes	yes		yes
Oriental bittersweet						yes			yes
plantain	yes	petroleum	Cu, Cr			yes	yes		yes

PLANT	EROSION CONTROL	REMOVES TOXIN	REMOVES HEAVY METALS	CLEANS WASTEWATER	CLEANS AIRBORNE POLLUTANTS	PROVIDES WILDLIFE FOOD	PROVIDES BEE NECTARY	ENRICHES SOIL	PROTECTS EXHAUSTED & TRAUMATIZED LAND
purple loosestrife	yes	nitrogen, phosphorus		yes			yes		yes
reed	yes	herbicides, petroleum, TNT, DDT, PCBs, xenobiotics, chlorobenzene, phenols, sulphide, acid orange 7, polycyclic aromatic hydrocarbons	Zn, Ph, P, Pb, Cd, Cr, Ni, Mn, Cu, Zn, Se, Na, Th, U, Cs-137	yes		yes			yes
Russian olive	yes					yes	yes	yes	yes
Scotch broom	yes	pesticide (HCH)	Mn			yes	yes	yes	yes
Siberian elm	yes	perchlorate	Fe, Mn, Al, Zn, Pb, Ni, Cr, As, Cd, Cu				yes		yes
tamarisk	yes		Cd, Pb, Cu, As, Na	yes		yes	yes		yes
thistles	yes					yes	yes		yes
tree-of-heaven	yes	yes	yes	yes	yes	yes	yes		yes
white mulberry						yes	yes	yes	
wild mustard	yes						yes		yes
wild rose	yes					yes	yes		yes

People are beginning to understand that if Achillea millefolium *can stop hemorrhaging and help the healing of wounds that are caused by technology it can, as well, help an oozing erosion gully that is breaking our hearts, or the scarred landscape created by a bulldozer. There is a reason they spring up in some logged forests. Such medicinal understandings can become quite subtle and complex. Combinations of plants, just like those in herbal tinctures, can be planted together to synergistically help damaged ecosystems. Herbalists can play a role in the healing of ecosystems, especially if they begin to understand the similarity between human organ systems and planetary ecosystems. Such folk and indigenous healers, because they tend to see the pattern that connects, because they have a functional sense of aesthetic unity, are important healers for the planet itself.*

STEPHEN HARROD BUHNER,
THE LOST LANGUAGE OF PLANTS

As we have now seen, in addition to protecting fragile, damaged eco-systems and enriching biodiversity throughout these areas, invasive plants help to clean from the environment many toxic overloads. The accumulation of heavy metals and poisonous influences within ecosystems is staggering, and because these toxins rarely produce any sight, smell, or sensation we can perceive with touch, the land and people remain poisoned. Examples abound to reveal the neglect of corporate and governmental action involving the use of toxic influences. (These, however, go beyond the scope of this book.) Because we fail to remember the repercussions of these pollutants and their inherent connection to human activity, opportunistic plants come to us as messengers to warn us of our follies. With time, these plants will fulfill their purpose; we will understand their significance, and the seeds of the extinct plants will find a fresh breath of becoming and will sprout from the organic humus of deep, dark patience.

Sometimes it takes awhile before we realize the amazing capabilities of what lies under our very noses—even old, smelly mold growing on stale bread and cheese turned into the disease-eliminating savior of the twentieth century: penicillin.

The planet will be here for a long, long, LONG time after we're gone, and it will heal itself, it will cleanse itself, 'cause that's what it does. It's a self-correcting system. The air and the water will recover, the earth will be renewed, and if it's true that plastic is not degradable, well, the planet will simply incorporate plastic into a new paradigm: the Earth plus plastic. The Earth doesn't share our prejudice towards plastic. Plastic came out of the Earth. The Earth probably sees plastic as just another one of its children. Could be the only reason the Earth allowed us to be spawned from it in the first place. It wanted plastic for itself. Didn't know how to make it. Needed us. Could be the answer to our age-old egocentric philosophical question, "Why are we here?" Plastic bleephole.

GEORGE CARLIN, "THE PLANET IS FINE"

8

The Chemistry
of Plant Medicine

Plants are all chemists, tirelessly assembling the molecules of the world.

GARY SNYDER, FOREWORD TO *PHARMAKO/POEIA*

PLANTS ARE MESSENGERS NOT only between humans and nature but also among one another. They communicate directly with each other and their environs by adding certain compounds to and removing them from the soil or air, which creates an interplay of continuous chemical chatter throughout the environment. Like letters in an alphabet, elements and compounds combine to form a meaningful language that is a unique expression from plants. Various chemical compositions are relayed throughout a plant's environment, and at times these can be felt, seen, smelled, and heard for miles around by an assortment of species, including plants, animals, insects, and microbes.

Different chemical compounds signify different messages, and some of these are released consistently over time, whereas others are an immediate response to temporary stimuli. Certain species such as black walnut trees emit repellant chemicals in order to dissuade other plants from coming near. These chemicals relay a persuasive deterrent: stay clear of the demarcated areas. In contrast to black walnut's antisocial tendencies, plants in the Legume family make compatible companion plantings because they fix nitrogen in the soil and thereby enhance fertility for other species. At times, we purposely plant nitrogen-fixing plants so that they will mingle their chemistries with soil microbes and thereby enrich the soil for other plants in the area. Other plants have similar capacities to nourish the soil by bringing forth deeply stored nutrients and minerals from within the earth. Trees and herbaceous plants have developed a host of constituents to help contend with pathogens and aid their own innate immune systems, and plants in turn provide these compounds for the health of the surrounding ecosystem and its inhabitants.

For hundreds of years, researchers have been isolating these same compounds exchanged among plants and their habitats in order to discover biochemicals that can address human disease. In fact, most pharmaceuticals are derived from plant-based sources. The sheer number of individual chemical components within a single plant can total in the hundreds or thousands, and science has only skimmed the surface with regard to understanding the complex interaction of these various chemicals—interactions both between the chemicals themselves and within the ecology of the human body. Yarrow, for example, has been found to contain more than one hundred biologically active compounds, including lactones, flavonoids, tannins, coumarins, saponins, vitamins, minerals, sterols, sugars, alkaloids, amino acids, and even a blue coloring found in the essential oil. Of these components, many have actions that are antimicrobial, while some are anti-inflammatory and others are homeostatic and analgesic. Humans will never be able to comprehend the fullest extent of the interactions among all the chemicals in plants—many are biologically active at one part per million, and others are active at one part per billion or even per trillion. The multitude of different combinations within plant chemistries interact in complex ways. Certain chemicals can amplify the actions of others to potentiate healing. In other instances, chemicals can offset the harsh, toxic qualities that a compound can impose. Soil, air, light, water, season, time of day, and dozens of other unknown variables that a plant may experience within its livable territory can alter the potency of chemicals and the presence of chemicals in various parts of the plant.[1] Every day, scientists are discovering new constituents within plants, and the search goes on and on . . .

We tread infinitesimal steps in the quest to discover the wholeness of a plant . . . and still we don't find it.

Scientists studying plant compounds isolate and potentiate them into powerful pharmaceutical derivatives. These drugs are engineered in ways far removed from their natural state—they are isolated from the complex matrix of the entire plant and are concentrated into artificially large and dangerous doses. For hundreds of thousands of years, humankind has developed relationships with plant medicines—and the very records of this have been written with paper and ink made from plants. Our human bodies have developed a kind of plant language; they have learned to assimilate plants into the healing process. Up until a hundred years ago, plant medicines made up our main pharmacy—and the recent propulsion into pharmaceutical-based health care has resulted in turmoil for our bodies. The corresponding evidence is clear that human bodies, in general, do not assimilate or for that matter appreciate the highly concentrated synthetic drugs that are so ubiquitous in health care today.

A communication is relayed . . . read the label and experience the side effects to get the message.

What we must remember regarding the individual virtues of a single constituent is that the whole of a plant does not add up to the sum of its chemicals. Every plant is a multifaceted, interdependent living force of potent individualized medicines. The seemingly toxic actions of an isolated compound are harmonized by the other chemicals within the whole plant, and other components can increase the potency of the primary compounds and direct them to specific areas of the human body. Plants that are considered herbal antibiotics not only have antibacterial properties, but they also address a wide range of microbial pathogens and supply the body with immune-enhancing compounds. Within a single plant, infinite interactions between innumerable chemicals are complex and astounding and beyond our complete understanding. To believe that we can improve upon a plant's healing properties and wisdom is nothing but hubris. The methodology of dissecting plants and studying them to see how they work leads to infinite combinations with thousands of variables, and we could continue studying plants for hundreds of years if we wish to know a fraction of their compounds in the world. Plant chemists, then, point out the next layer of variables within individual compounds—those with subtle atomic shifts that continue to evolve and change. At times the plant seems cognizant of these shifts—the plant responds to the environment, its stressors, and the environment's inhabitants.

> *Plants have the ability to produce an almost endless number of chemical variations on a single chemical structure.*
> DAVID HOFFMAN, *PHYTOCHEMISTRY: MOLECULAR VERIDITAS*[2]

PLANT COMMUNICATION

Plants speak to their environment with the chemicals they release. They communicate to the microorganisms through the plants' root hairs deep in the soil, to those foraging insects that eat their leaves, and to the scavenging animals that eat their fruit. In addition, the plants and their chemicals convey messages for humans, and they speak eloquently with a multisensorial language that we have long admired. We can see this in the beauty of a flower, smell it in a plant's fragrance, and taste it in a plant's flavor—and we can feel it by the impact of a plant's medicine in our body.

. . . And then there's listening.

Plants emit chemicals into their environment to attract pollinators, repel pests, nourish the soil, and heal themselves and others. They are known to change their form depending on the growing conditions and alter their chemical signals if they feel stressed or threatened *or* if they feel affection or gratitude. Throughout time, many people have intuitively recognized this, observing plant changes in the wild and in their own gardens. In addition, the 1973 book *The Secret Life of Plants* highlighted the plant experiments of plant geniuses Luther Burbank (1849–1926) and George Washington Carver (ca. 1864–1943). Some studies cited in the book explored plants and how their growth and health responded to various kinds of music or different thoughts, feelings, and intentions. At the time of their work, Burbank and Carver were seen as going out on a limb, and today their names are fading, though their efforts to understand a plant's life are quite influential in our present lives. In fact, Luther Burbank was instrumental in the introduction and creation of many of our food plants, most notably the Burbank potato, as well as new forms of tomatoes, corn, squash, peas, and asparagus. He also successfully coaxed the genetics of a cactus to remove the cactus's spines and formed a wild plant into the famously adaptable garden flower Shasta daisy. George Washington Carver expanded the growing abilities of sweet potatoes and soybeans into southern agriculture and brought to light the peanut that revolutionized our society with its butter, tubers, and oil. This single plant has sustained several generations and has created an American food culture, complete with peanut butter and jelly sandwiches and roasted nuts for the baseball tradition. With the help of Burbank, he was also responsible for deep-frying potatoes—that is, creating french fries.

Carver and Burbank worked with plants in ways that not only changed the chemistry within the plants but also dramatically changed their form. To accomplish this, these plant geniuses used a listening device within their hearts and melded with the plants in order to unlock their secret lives and innate potentials that stretch back into the species' deep genetic ancestry. The earth's laboratory is always at play, always changing, and always striving for an unachievable equilibrium, which means that plants and their compounds interact with and adapt to countless factors.

> *The secrets are in the plants. To elicit them you have to love them enough.*
>
> GEORGE WASHINGTON CARVER

Imagine a tree that constantly serves as forage for hungry animals. One point of view indicates that it would make sense to fence off this tree if we want it to flourish. Yet nature challenges our initial impulse on the matter with examples of surprising interspecies links and dependencies. The journal *Science* published a study in which researchers discovered a mutually enhancing relationship between an ant species and African acacia trees. The aggressive, biting ants protect the trees from foraging elephants, giraffes, and other animals, and in return, the trees provide shelter for the ants and nectar for their food. In the study they fenced off some of these trees to determine how they would respond without the browsing herbivores. After a few years, the fenced acacias began to look sickly and did not grow as well as the unfenced trees. The scientists discovered that the trees stopped taking care of the ants because the creatures were not needed for protection anymore. The ants then began to damage the acacias, or they moved from them, which allowed other ants and insects to move in, and these new creatures sickened the trees. As Ted R. Schultz, a researcher from the Smithsonian Institution's National Museum of Natural History observed, "The system reported here is a balance of a number of players—the trees, the browsing mammals, the main ant and three other ant species, with the ants all competing for the trees. Remove one of the players—the browsing mammals—and all the other moving parts rearrange themselves in a way we hardly could have predicted."[3]

One foraging animal sent to the tree a message of sorts: "I won't be eating your leaves, so take the easy life," so the tree stopped producing any immune-responsive chemicals. The plant then sent a message to the symbiotic ants and stopped providing nectar and the ants' essential sugar compounds: "I don't need you ants anymore." The ants countered with their own fight-or-flight adrenal responinse with "I will attack you" or "I will leave you." At last, the plant, like a diabetic who needs insulin, sends out a chemically imbalanced message. But the tree needs a bite on a daily basis and cries for life—"Someone just eat my leaves"—in order to produce the necessary compounds for survival.

> *Chemical communication between herbivores and plants is the primary driving force behind what we recognize as coevolution, and that process is better represented as an accelerated process of genetic change facilitated by the efficiency of chemical information exchange.*
>
> KEVIN SPENCER, "THE CHEMISTRY OF COEVOLUTION"

Scientists have now discovered ways to "listen" to the sound of plants in order to understand them and their environments. In analyzing the properties of a plant's leaves, researchers in one study established communication by ultrasound with above-audible frequencies to determine water content, thickness, density, or compressibility. Researchers start by emitting broadband ultrasonic pulses (between 0.2 and 2 megahertz) directly on the leaves, which in turn begin vibrating the leaf. As the vibrations ripple off the leaf, an ultrasonic sensor detects the waves. These signals form a distinct resonance of the leaf, which are analyzed and converted into terms to assess the leaves' characteristics. According to Tomas E. Gómez, a researcher in the study, "The method involves establishing a silent dialogue with plant leaves, questioning them and listening to what they say. . . . The voice of the leaves itself is what gives us information about their status and their properties, all in an innocuous and silent way."[4]

In another study to detect and measure contaminants in a body of water, scientists are using microscopic algae species and have developed methods to determine the photosynthesizing potential of these plants, which then serve as a measure of the health of the environment. Scientists use green lasers to test the ability of the algae to photosynthesize such light and to see how the plants might cope with toxic surroundings. Plants convert the greater potential of photosynthesis into energy, which then creates heat. Heat expands and changes the pressure within the water, and the corresponding pressures emitted from the plants produce sound waves that then can be transmitted through special microphones and heard as a specific resonance. As algae were subjected to toxins, there was a change to the quantity of light energy that was converted into sound, and scientists have learned to interpret this language to determine contamination levels. For example, according to researcher Yulia Pinchasov, "Algae suffering from lead poisoning, like waste discharged from battery and paint manufacturing plants, will produce a different sound than those suffering from lack of iron or exposure to other toxins."[5]

Plant communications are like stones in water. The ripples they create move throughout ecosystems; they wash up against us. That we take plant words in through our nose or our skin or our eyes or our tongue instead of our ears does not make their language less subtle, or sophisticated, or less filled with meaning. As the soul of a human being can never be understood from its chemistry or grammar, so cannot plant purpose, intelligence, or soul. Plants are much more than the sum of their parts. And they have been talking to us a long time.

STEPHEN HARROD BUHNER,
THE LOST LANGUAGE OF PLANTS

This way of direct communication with plants has been known for ages by the native-minded people of the world. Intimately calling and singing to a plant when it was needed for medicine allowed an individual plant to prepare for the healing and also potentiated its medicine. Speaking the plant's language in this way allowed the initiator to become an impetus to changing the chemical structure of the plant. The healer projects with conviction loving, gentle feelings, and the plant then responds with caring and healing compounds. The vibrations that people exude directly impact the elements of the universe, and this is felt by all of us, all of the time. Although, we don't always choose to feel.

Just imagine the difference between what we feel when an angry person walks into the room versus the entrance of someone who is happy.

A flood of adrenaline or—in case of the latter—a cascade of endorphins is released. Just as the presence of other people has the power to affect our experience, so too the presence of plants can alter our mood, feelings, and behaviors. We can begin by noting a plant's emotional undertones, but the complex communications they transmit do indeed go deeper. As we attune to a plant, we may begin to feel our stomach tighten, or we may have a flash of lightheadedness—all are triggered by the plant and the chemistry it possesses. We can sift through the layers of information embedded in the communications of the plant and find a multitude of responses and feeling tones transmitted by the plant. These are the same vibrations that can be read by the ultrasonic sensors and underwater microphones. The vibrations are subtle and rarely audible; they rely on our own innate heart sensor to read the relay of vibrational messages. We may have flashes of intuition— uses for medicine, knowledge of ecological interactions, and other purposes the plant may have for growing where it does. An unfolding process of deepening understanding of the plant is *phytognosis*. We then hold on to these stories of the plant and, as Stephen Buhner says, "wrap them up in your heart-cloth."[6] In order to further understand their meanings, we bring out these experiences and then share these medicine stories with others. This ability is not unique to some but is shared among us all through an inherited gene long used throughout our ancestry and available to us whenever we choose to use this gift of communication.

Walking into a disturbed environment with some of the overbearing, thorny, and expansive invasives, some people might feel angry, which leaves them with a bitter taste in their mouths about the spread of these invasives. These first impressions are important clues as to the impact the plant has on us through chemical information relayed invisibly. They stimulate a process within us, and they can have multiple effects on different people, just as plant medicines stimulate different bodily systems.

> As we settle into the feeling of the plant's presence and allow
> our rationalizing brain to relax, full immersion with the plant
> occurs, and we achieve a knowing of the heart.

The medicine is the response in us, and the anger that we may feel is the gateway to releasing the bile that has bound our ecologic body—and the plant guides the process. Sometimes we feel guilt or heartbreak, and the depth of tension within us is reflexively released by the added stimulant of the plant chemistry that facilitates the emotion's liberation. This is a kind of therapy—but instead of a human therapist who relates to our psychosomatic landscapes and who points to and touches those places where we may be mentally or emotionally stuck, there is a plant therapist.

It is not half so important to know as to feel.

RACHEL CARSON

People have had relations with plants as long as they have had relations with one another. Why, then, should our communication with them be any different from talking with people? All traditions have had ways of receiving information from plants, and some of us still follow our ancestors and directly, imtimately connect to the green world. May we all continue to dream with the plants, receive their medicine, and share in their illumined stories, just as people have done by the light of fire for millions of twilights.

> *Many scientists have remarked with surprise that Luther Burbank, George Washington Carver, and even the Nobel laureate Barbara McClintock all have said that it was the plants who told them what to do, who revealed their mysteries to them. The only requirement, they commented, was that they had to care for them, to treat them with respect, to have a feeling for the organism. This would not be strange to the Winnebago, among whom it has been said that people must treat the plants like human beings, make proper offerings, and treat them with respect if they wish their help. Nor would this be strange to a four year old sitting and talking with flowers.*
>
> STEPHEN HARROD BUHNER,
> *THE LOST LANGUAGE OF PLANTS*

PLANT MEDICINE

Herbalism is based on relationship—relationship between plant and human, plant and planet, human and planet. Using herbs in the healing process means taking part in an ecological cycle. This offers us the opportunity consciously to be present in the living, vital world of which we are part; to invite wholeness and our world into our lives through awareness of the remedies being used. The herbs can link us into the broader context of planetary wholeness, so that whilst they are doing their physiological/medical job, we can do ours and build an awareness of the links and mutual relationships.

WENDELL BERRY[7]

We have been linked to the healing plants around us for hundreds of thousands of years. All cultures and traditions throughout history have been reliant on them, and it is impossible for us to have reached where we are—this great global civilization—if we had been using placebo-based, noneffective plant superstition for the past one hundred thousand years or, for that matter, the past two hundred. All the billions of people throughout time who have been using plants as medicine could not have all been so awfully wrong. As David Hoffmann points out, "the history of herbalism is the history of humanity itself,"[8] and all people are immortally linked to this ancient way of learning from and using plants.

Although, these days the disconnection between the natural world and humans has grown so deep that plant medicines have become disregarded as old-fashioned, unscientific, ineffective, and dangerous. When any herbal result is demonstrated, it is most often discounted as no more than the effect of placebo. Most doctors in today's world fail to recognize the molecular plant origin of most of the pharmaceuticals they prescribe. Appreciating plant medicines was commonplace among doctors a hundred years ago, and having a familiarity with plant botany and chemistry and with the herbal (animal and mineral) preparations they prescribed was a prerequisite for the job. Even the science of pharmacology, with its intimate origins in plants themselves, has distanced itself from the natural world: medical colleges have dropped courses in pharmacognosy, the "branch of pharmacy that dealt with the discovery, extraction, and identification of new drugs from natural sources." This was the "living heart of pharmacy and that field's connection to nature,"[9] the botanist doctor Andrew Weil conveys—and until Dr. Weil came along, there had been no botanical medicine focus in any medical doctor training for at least the past sixty years.

There are people who deal with patients who have not the slightest interest in or knowledge of botany. I think that's sad. I feel lonely in my position, and sad because it shows the degree to which science and medicine have separated themselves from nature and separated us from nature.

ANDREW WEIL, M.D.[10]

The standard Western medical practice of dispensing pharmaceuticals and demonizing plants for illness runs contrary to practice in the rest of the world, where, according to the World Health Organization (WHO), 80 percent of the global population relies on traditional herbal medicine and therapies. The WHO has recognized the importance of plants in the health and healing of humans and has declared them "the people's medicine"—for plant medicines are available free of charge to all those in the world who are ill. Because our culture has been ingrained with the response that plant medicines are unscientific, we often have been blind to the very remedies we need to cure the present-day diseases that have been amplified by the failing practices of modern medicine. In general, herbal medicines have been deemed ineffectual based not on studies that disprove their effects but on the fact that *there are no studies that prove their effects*. Herbs are declared dangerous because of a few exaggerated claims of harm resulting from inappropriate use—yet before a pharmaceutical is removed from the shelf, thousands of people may have already died, and hundreds of people are continually poisoned to fatality from aspirin, Tylenol, and non-steroidal anti-inflammatory drugs (NSAIDs) every year (at least 117 in 2008). Meanwhile, not one death was attributed to the use of a plant-based medicine in 2008.[11]

"The irony is that according to medicine's own standards, 85 percent of prescribed standard medical treatments lack scientific validation," says Kenny Ausubel, founder of Bioneers,[12] and according to a 2000 study in the *Journal of the American Medical Association* (*JAMA*), at least two hundred twenty-five thousand deaths can be attributed every year to the unintended consequences of medical procedures and treatments that are considered conventionally acceptable.[13] In addition, more than one hundred thousand deaths annually are due to "nonerror adverse events of medications," with unknown risk factors to blame; eighty thousand people die from hospital stays; and another forty thousand die from unnecessary surgery, medication errors, and other hospital mishaps. All these statistics add up to a huge percentage of the population—leading to the conclusion that conventional medical treatment is a leading cause of death in the United States.[14]

> *It is becoming evident that in order to survive the multifaceted crisis at hand, humanity must learn some environmental humility, including how to cooperate with nature. Herbalism is a unique and important expression of this cooperation.*
> DAVID HOFFMANN, *MEDICAL HERBALISM*

When seen in relation to a handful of deaths resulting from the improper use of herbs, the modern standard of safe and reliable medicine begins to look illogical and unsubstantiated at best. The documentation for modern pharmaceutical uses spans the past twenty-five to fifty years, while plant medicine has been tested by humans for hundreds of thousands of years. The ancient medicinal arts from traditional Chinese medicine, Western herbology, and ayuvedic traditions remain intact today with more than two thousand years of written records. In other parts of the world, oral tradition has kept plant medicine knowledge alive, strong, and expanding. Within certain cultures, the first missionaries finally documented instructions for how to use the plants for healing. Nevertheless, many accounts of healing plants were not comprehended by analytic minds and were misconstrued by colonizers due to language barriers, spiritual beliefs, and the colonizers' underlying judgment of indigenous people. Though the modern scientific community discredits plant medicines as being substantiated only by mere anecdotal evidence, the conventional medicine practiced today employs its own uses of pharmaceuticals based on nothing but subjective data tested by scientists who are on the payroll of the pharmaceutical companies themselves, then dispensed by doctors who often receive financial incentives for prescribing them.

The vested interest of the pharmaceutical companies negates the validity of the findings. The very foundation of our modern medical system is itself unhealthy; it is nothing more than big business predicated on and fed by illness rather than wellness. A shaman has no vested financial interest in the type of medicine a patient takes as long as the malady improves, but a pharmaceutical company makes little profit in cases of complete recovery— yet stands to gain much through lifelong prescriptions.

Plants are indeed the people's medicine, and they grow all around us. The earth forms them, and they cannot be patented and capitalized upon—unlike the drugs that pharmaceutical factories pump out, those that so often have unintended toxic side effects. Plants grow year after year, providing a renewing medicine cabinet, and humans have used them in this way for a long time. This knowledge is available to all people who want to know and cannot be manipulated by those who desire power over anyone's health and healing. In order to deal with the chronic illnesses, infectious diseases, and health crises of today, we must return to our root doctors and see the healing power in the sprouting green pharmacy all around us. In chapter 9 we will discover the potent medicines of the wild and rambunctiously spreading plants.

For, with their abundance, the plants are asking to be of service.

Seeking for truth I considered within myself that if there were no teachers of medicine in this world, how would I set to learn the art. Not otherwise than in the great book of nature, written with the finger of God. I am accused and denounced for not having entered in at the right door of the art. But which is the right one? Galen, Avicenna, Mesue, Rhais, or honest nature? Through this last door I entered, and the light of nature, and no apothecary's lamp directed me on my way.

PARACELSUS

9

Using Invasive Plants to
Treat Invasive Diseases

*Use of invasive species of plants in treatment reduces the
impact on non-invasive medicinals and begins using plants
that are accompanying invading pathogens as they move into
new ecoregions.*

STEPHEN HARROD BUHNER, *HEALING LYME*

AMONG THE MANY ACCUSATIONS against invasive plants, one of
the most intriguing and misleading is the idea that they are responsible
for the spread of invasive, infectious disease. Finding greater densities
of Lyme disease–bearing ticks within stands of invasive Japanese honeysuckle and barberry than in surrounding areas, the Maine Medical
Research Center concluded these plants promote the spread of this epidemic disease. These hasty conclusions remind us that correlation does
not necessarily confirm causality. Explored within a different context, the
proven correlation between migrations of Lyme disease–bearing ticks and
Japanese honeysuckle and barberry suggests that the plants contain cures
to this disease. The differing conclusions from the same study represent
an example in which an adjustment of perspective is essential to understanding complex interactions within a conscious ecosystem that comprises conscious species. As invasive plants and epidemic diseases enter
our lives at similarly disturbing rates, we are left to ponder the nature of

the relationship between the plants and disease. Before we continue to wage war against the plants in the name of warding off diseases, we must consider the possibility that we are destroying potent medicinal remedies to the very diseases against which we want to protect ourselves.

The idea of using these invasive plants to treat invasive, infectious diseases was presented to me by Stephen Buhner. In his book *Healing Lyme,* Buhner brings to light the medicinally significant connection between pandemic plants and diseases by stating, "invasive plant species are specifically indicated for use with invasive or emerging diseases such as Lyme, West Nile encephalitis, SARS, hepatitis C, HIV, and so on."[1] While researching plant candidates for treating Lyme disease, Buhner came upon the use of Japanese knotweed as a potent antimicrobial, antioxidant, and

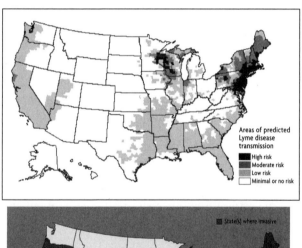

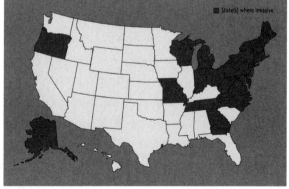

Maps comparing the spread of Japanese knotweed to the spread of Lyme disease. The Japanese knotweed distribution map is from the National Park Service, www.nps .gov/plants/alien/fact/faja1.htm, and the National Lyme disease risk map is from the Centers for Disease Control, www.cdc.gov/ncidod/dvbid/lyme/riskmap.htm.

immune-enhancing medicine. He then realized that knotweed has spread throughout North America in nearly the same trajectory and at the same rate as Lyme disease, with both exotic infestations creeping in during the past ten to fifteen years and creating today's explosion. Within these widespread occurrences, the plants that offer themselves and their gifts of healing for rampant epidemic diseases and antibiotic-resistant infections also offer hidden benefits.

They offer a kind of homeopathic medicine: "like cures like."

THE END OF AN ANTIBIOTIC WORLD

The fervor of using antibiotics began nearly seventy years ago, and since then, all foundational life-forms have felt the deep repercussions of that decision. Myopic medicine and science has fed antibiotics—meaning literally "against life"—to people and the environment by the billions of pounds in order to kill off microorganisms. Hospitals alone have an artillery of antibiotics, and outside the hospital setting, physicians dispense more than two hundred million antibiotic prescriptions. Overuse of antibiotics is not only directly detrimental to human health—flooding our bodies with concentrated antilife compounds, compromising our immune and digestive systems—but it also contributes to the rise of strains of resistant bacteria: astounding superbugs that have adapted and evolved to survive the onslaught of our state-of-the-art weaponry.

Within years of the advent of penicillin in 1942, strains of *Staphylococcus* bacteria were resistant to the antibiotic, and by the 1970s, outbreaks of antibiotic-resistant staph swept through hospitals all over the world, killing many patients along the way. Since then, bacteria and other microbes have adapted to the constant assault on their kind by using strategic evolutionary measures. Bacteria have learned to alter the internal structures of intended targets, and they can vary their rate of assimilation by changing membrane structure and permeability. These resistant microorganisms have also learned to create compounds that destroy antibiotics, and in some novel cases of adaptation, they have learned to use these chemicals as food. In addition, the innate intelligence in the microbial world means that bacteria communicate and pass information to their offspring very quickly by creating new generations every twenty minutes or so (about five hundred thousand times faster than humans), and they relay to other bacteria the mechanisms that have helped them adjust to antibiotics.

The expertise in the 1940s and '50s, including the talents of Nobel prize–winning scientists and a U.S. Surgeon General, declared the "virtual elimination of infectious diseases" and were predicting tuberculosis and malaria to be erradicated by the year 2000.[2]

WHERE WE ARE TODAY

As I write this in the winter of 2008–09, the media reports that 99 percent of all flu viruses in the United States this season are resistant to the antiviral drug Tamiflu, and other infectious diseases continue to spread rampantly. Although spontaneous mutations within the flu virus resulting in this drug resistance has shocked and puzzled scientists, it should come as no surprise. We need only look at the history of microbial warfare that has resulted in multidrug-resistant strains of *Mycobacterium tuberculosis, Staphylococcus aureus, Streptococcus pneumonia, Pseudomonas aeruginosa, Streptococcus pyogene, Enterococcus faecium, E. coli,* and *salmonella,* among others. Throughout the United States in 2009, more than sixty-five thousand people were killed by drug-resistant infections—more than prostate and breast cancer combined—and over nineteen thousand people died from a staph infection alone.[3]

> *This situation raises the staggering possibility that a time will come when antibiotics as a mode of therapy will be only a fact of historic interest.*
>
> DR. STUART LEVY[4]

We are witnessing a time of collapse in our health and in the efficacy and safety of the care people routinely receive. Experts have failed to heed the fact that antibiotics are creating stronger pathogens and weakening our immune systems, and the repercussions manifest in more epidemics and greater rates of chronic disease. We have advanced technology, yet we have created more disease. We have embraced the precision of microscopic vision to dissect single cells, yet the interdependent systems of the entire body have all but been forgotten. Doctors and institutions focus on disease care instead of health care, and modern medicine has excelled in the creation of pharmaceutical treatments of the symptoms of disease rather than in treating the underlying cause of illness. This has necessitated more drugs in the ongoing management of side effects.

Fortunately some doctors are beginning to step out of their box, changing the common practice of widely dispensing antibiotics. A study from Norway tells of a simple solution to stop a killer superbug: stop taking so many drugs. About twenty-five years ago Norway implemented measures to combat methicillin-resistant *Staphylococcus aureus* (MRSA) that spreads throughout hospitals, killing many along the way. Their program has made Norway the most infection-free country in the world today. One of the most telling factors for this low insidence was that the doctors prescribe far fewer antibiotics, and therefore, there is less chance of resistance to buildup. This approach has been replicated at five hospitals in the U.K., where at one hospital in five years, the MRSA rate dropped from 47 infections to just one. According to Oslo's MRSA medical adviser, Jan Hendrik-Binder, "It's a very sad situation that in some places so many are dying from this, because we have shown here in Norway that methicillin-resistant *Staphylococcus aureus* (MRSA) can be controlled, and with not too much effort."[5]

Despite this knowledge, MRSA continues: in the United States it kills nineteen thousand people a year, while people consume more than 50 million pounds of antibiotics every year. Many more expired drugs are alternatively flushed down the drain or sent to landfills. Our livestock are fed more than 20 million pounds, hospitals dump at least 250 million pounds of active pharmaceuticals into public sewage systems, and pharmaceutical manufacturers legally release at least 271 million pounds of drugs—all of which seeps into the earth and waterways. With modern medical practices, we are experiencing a worldwide epidemic of pharmaceutical contamination in lakes, rivers, and reservoirs and in the seemingly pristine waters of the high Alps and the North Sea. These contaminations affect all life on Earth.[6] As with all pharmaceuticals, after partial assimilation by the body, antibiotics are excreted unchanged and remain active. They are then flushed into septic systems and down sewers and into the environment, affecting all wild bacteria with which they come in contact and helping them develop resistence. Even for the conscientious objector who does not take them directly, consuming antibiotics has become unavoidable, because they have contaminated our soil, drinking water, and food supply—from the glass of water we pour from the tap to the burger on the bun, antibiotics are ubiquitous and inescapable. The abuse of antibiotics is leading to increasing rates of food pandemics that contaminate grocery store shelves and consumer kitchens with *Salmonella, E. coli, Shigella, Listeria,* and the like.

We are witnessing incomprehensible evolutionary changes within these antibiotic-resistant microorganisms, as evidence by the emergence of superbugs and other unique pathogens that scientists fear could lead to epidemics of infectious disease that are more potent and deadly than any yet confronted in history.

With every action, there is a reaction . . . cycle after cycle.

In healing traditions throughout the world, human beings are seen as microcosms within the greater macrocosm around us. These larger and smaller expressions of the world mirror and relate to each other, and superior physicians meld both the microcosm and the macrocosm in their search for the underlying cause of illness. Traditional Chinese medicine (TCM) explains how to find balance of the yin and yang and to follow the laws of nature in order to resist disease. The Western eclectic physicians of old believed "as above, so below" and used this logic to understand how the heavens (that is, the outside world) influenced health and disease (that is, our interior world). Modern psychology (and common sense) tells us that our interior state can be affected by our environment, and at times our inner states of consciousness project onto our outer lives and the circumstances around us. The traumatized and poisoned landscapes in which some of us live can become like a deep wound that impacts health in many subtle ways, sometimes manifesting itself with debilitating physical ailments and serious internal disease.

For a long time now walls have been built around human health (in the form of penicillin, vaccines, etc.). Our separation from nature has begun to leave an emptiness in us that must be filled. Today's widespread afflictions are like the spreading cracks in a dam: the overgrazed, barren land and the traumatized, subdivided terrain that must be tended to and stabilized in order to support a healthy, living bodily system. Infectious diseases represent this crack in the dam, and the dam is about to burst. The more we till our soils with antibiotics and pharmaceuticals, the stronger diseases become, and the more we pollute our waters, the weaker are our innate healing responses. In disturbed and toxic areas of the body, the opportunistic pathogens can abound and create more sickness. People, like the earth itself, are overburdened with toxic and infectious burdens, and all ecosystems cope with poisoned and traumatized influences. Once environmentally degraded areas are left to nature, however, a unique plant community breathes life into the desolate land and heals the fractures in the ecosystem. These opportunistic, expansive plants also can do this for human health and well-being.

> *The collapse of our surrounding native ecosystems and life-forms due to human desire is another reflection of our severance from the natural world and mirrors the loss of our own native self. We try to hold on to an idea of nativity in the outer world, because we have lost it inside ourselves. The more we begin to fill in the cracks and recognize the healing that comes from the wildness around us, the more we recognize the work of the ever-changing cycles of natural ecosystems. The more we are present to the continually evolving, self-creating, healing wisdom of nature, the more we begin to see the dynamic interplay of all forces intermingling. As the forces become more coherent and interlaced, we are enveloped by an insight into the grand mystery that never rests. Inner worlds and outer worlds commingle— and an understanding of one leads to greater understanding of the other. Thereby, we achieve a return to our native self.*

NATURE'S ANTIBIOTICS

Unlike the unilateral approach of pharmaceutical antibiotics, which afford ample opportunity for bacteria to build resistance, the comprehensive wisdom of nature found in whole plant medicines renders microorganism resistance impossible. The multifaceted chemistry inherent in plants is too complex for bacteria to develop immunity. As we have already discovered, plants contain dozens, if not thousands, of chemicals that interact with harmonizing, enhancing, and even antagonizing effects, each of which impacts the ecology of the body. When we ingest just one plant as medicine, a formula of potentially hundreds of components interact with the pathogen, making it very difficult for it to develop resistance or to avoid the effect of the medicine. The same bacteria that find it easy to adapt to the single chemical dynamic within conventional antibiotics are impotent against the intricate chemical matrices of whole plant medicine.

It takes bacteria a few generations (hours to days) to gather resources and devise adaptations to the relatively simple, singular component within an antibiotic, and modern medicine has begun to see the peril of these pharmaceuticals, so now doctors commonly administer a formula of antibiotics. Though this is an obvious step to increase the efficacy of pharmaceuticals, the multidrug medicines still fail to follow nature's intrinsic formula: plant compounds mutually enhance antimicrobial effects, direct the actions to afflicted areas of the body, strengthen innate immune responses, and mitigate potential toxicity of certain compounds.

Today's pharmaceutical antibiotics are quite bitter and similar in nose-turning bitterness to their plant predecessors. The bitter flavor generates a wide range of responses in various systems of the body, exemplified in the numerous cure-all bitter plants and potions used throughout history. The *bitter principle* produced by many plants initiates basic healing responses by increasing bile flow and stimulating the detoxification processes in the liver, which in turn helps relieve the body's tissues and blood of infection and toxins. These bitter alkaloids help regulate digestive hormones, aid in the metabolism of foods, and promote the self-repair mechanisms of the stomach and intestines. Many invasive plants possess a very bitter component, exhibiting antibiotic or antitoxin properties that can kill off a broad spectrum of pathogens and parasites. One of the most notoriously intense bitters is wormwood (*Artemesia absinthium*), and Stephen Buhner describes it as "one of the most powerful herbs for the treatment of antibiotic-resistant disease available."[7]

> Wormwood is not just available—it grows widely, and some consider it annoying.

Many of the diseases that have invaded our landscape and evaded technological medicine are described in language that parallels that used to describe invasive plants and the language of conventional warfare used to descibe disease treatments. Invasive diseases are caused by opportunistic pathogens that are growing ever resistant to antibiotic pesticides. Coinfections of multidrug-resistant pathogens are a serious concern, especially in hospitals, and all the time we hear alarm about the next epidemic: AIDS, West Nile virus, avian-swine flu, SARS, Lyme disease . . .

> At the same time, plants bearing cures for these diseases are invading your backyard.

Still, doctors and hospitals continue to stockpile large amounts of antimicrobial drugs to combat the coming pandemic. Fortunately, as you will see, many of the invasive plants we humans trample on are trying to get our attention.

INDIVIDUAL INVASIVE DISEASES

Lyme Borreliosis

This is one emergent and invasive disease that I know intimately, both as a practitioner and as an infected individual. Most of us now know someone who has been infected with Lyme disease, and we may also know of the controversy that overshadows it. There is a vast array of conflicting opinions: some doctors claim there is no such thing as chronic Lyme disease, and others claim that 75 percent of the population is already infected with it. Myths prevail in the Lyme wars, but the shroud is beginning to be lifted. The most common conventional description is that a simple deer tick leaves a bull's-eye rash on its bite victim, the doctor prescribes some antibiotics, and a blood test confirms whether the victim is still sick.

"Don't worry. It's a rare disease, and the fact that you are tired and achy all the time just means you are getting old."

Yet much is overlooked in this conventional infection-and-treatment program: deer ticks are not the only host, a bull's-eye rash occurs in only about one-third of cases, testing for antibodies of Lyme spirochetes in the blood is notably unreliable, and antibiotics are most effective if administered in acute infection though there are numerous individuals who relapse. Let's sift the facts from the mountain of myths.

Lyme Borrelia spirochetes are ancient organisms, though in present form they first emerged in the 1970s, and Lyme disease was identified as the cause of strange cases of juvenile arthritis in Lyme, Connecticut. Since then, the disease has expanded far from the Connecticut shore—throughout the Northeast—and has widened its scope throughout North America and the world, aided by animal migration and human travel. The Centers for Disease Control (CDC) still continue to downplay the epidemic, claiming a fraction of the numbers of infected maintained by reputable sources such as Yale and Harvard University. The Ivy League studies affirm upward of two hundred thousand to three hundred thousand new cases each year, which, since the 1980s, adds up to infection of potentially one-quarter of the population. Many doctors remain ignorant of the disease, and they continually misdiagnose it or insult patients with the "it's all in your head" diagnosis. The only treatment they know is antibiotics, however minimal the drug's success unless administered in the early stage of infection.

The deer tick is the most common vector for the spread of this disease, but evidence presents other transmitting hosts of the Lyme spirochetes. Live Borrelia spirochetes have been found in other biting insects (various ticks, mosquitos, fleas) and in human bodily fluids (tears, saliva, urine, semen, breast milk). There have been confirmed instances of a mother passing infection to a baby in the womb. The CDC found that blood transfusions are another mode of transmission, because the Lyme bacteria survived the standard processing procedures normally applied to blood. A study by the University of Wisconson found that Lyme disease could also originate as a food-based infection. This alarming finding was derived from the fact that dairy cattle and other food animals can become infected and thus transfer the disease in raw foods and drinks.

The bacterium is difficult to study because the Borrelia organism cannot survive in a laboratory setting. What is known is that the spirochetes like to hide out in viscous areas of the body, most notably the joints, eyes, and central nervous system (brain and spinal cord), ultimately penetrating and affecting every organ system. After the disease is contracted, the organism can lie dormant for years or begin to wreak havoc immediately. For some people, the disease goes unnoticed as it slowly invades the body, hidden from the immune system, waiting for an opportune time to come out and flourish. For others, it penetrates deeply and quickly, creating a chronic condition nearly overnight.

This was my personal experience. I pulled off the tick one day, had intense flulike symptoms two days later, then two days after that I had severe neck pain. I woke up on the sixth day after I had been bitten, and I felt as though I had aged fifty years overnight: I experienced severe fatigue, weakness, and body aches. It so happens that the disease caught me under stress, with a weakened immune system. I was overextending myself, running a business and practice, raising a one-year-old daughter, and building a house. I bypassed the antibiotic route and started taking an herbal regime. I felt better fairly quickly, regaining my strength after three months. At this point, I mistakenly discontinued the herbs, and in three more months, I relapsed. I became severely ill with a cold and flu throughout the winter. I spent many sick days in bed and experienced achy joints, fatigue, weight loss, eye sensitivity, depression, and irritability. My health deteriorated to the point where I had to take time off from work. I experienced virtually the whole gamut of symptoms many people experience, and only slowly did I regain my strength and stability. I learned firsthand the tenacity and sneakiness of these bugs. While I took herbs the first three months, they had simply retreated to the corners of my body and waited patiently for a time to come out again. After this experience, however, I had learned my lesson and continued the herbal protocol steadfastly, four times daily for the next twelve months. Only in the eleventh month did my stamina return enough to allow me to walk the hill of my driveway without having to rest afterward. My recovery was a slow process, and it took two months to withdraw from the herbs in order to ensure my health. To date, I've had no relapse, but because I continue to live in an epidemic area, I am always on the alert for signs of the infection not only in myself but also in my family and patients.

Years before my own experience with the disease, I treated people known to be infected with Lyme disease and others who displayed symptoms but had no confirming blood test. Many people who come to my practice have been down the traditional route of ongoing courses of antibiotics—including, for some, receiving IV treatment—without significant improvement. Others have experienced a roller-coaster treatment given by their primary care physicians and specialists. They have submitted to various tests and received various diagnoses, and the ultimate conclusion is IAIYA (It's All in Your Head) disorder. These people fell through the cracks, and often in my practice I have the job of janitor, sweeping up the pieces and trying to restore some integrity and health to these people and provide a sense of hope for those who have been broken after years of experimental treatment and testing. Over the years, at least seven out of ten patients who remain loyal to the herbal protocol and adjunct therapies have seen substantial improvement in their condition. The core herb used in Lyme treatment is the wildly expansive plant Japanese knotweed.

Stephen Buhner has keenly observed that Japanese knotweed begins to inhabit locations six months to one year prior to the arrival of Lyme disease. Besides the use of Japanese knotweed, other herbalists have found successful treatment of Lyme disease by using other invasive medicinals. Matthew Wood uses teasel root and David Dalton uses the essence made from the flowers of teasel, a prolific plant in the upper Midwest. Dr. Zhang in New York uses Chinese coptis, which contains the powerful antimicrobial component berberine, also found in abundance in barberry, the purported nemesis of New England forests and an alleged promoter of Lyme disease. Other invasives deserve further study for use in treatment of Lyme disease.[8]

Invasive Plants Used to Treat Lyme Disease

- Japanese knotweed
- barberry
- dandelion
- teasel
- Japanese honeysuckle
- plantain

Mosquito-Borne Diseases

The devious mosquito is the pollinator of germs throughout the world, infecting people with West Nile disease; malaria; Rift Valley, dengue, yellow, and chikungunya fevers; and other deadly diseases. Some scientists predict that if the global warming trend we are experiencing continues, mosquitoes will flourish and the spread of diseases found commonly only in tropical climates will continue to move farther north into North America and Europe.

The virus that creates West Nile disease originated in the swamps of Africa and is related to the one that inflicts dengue fever. It is generally spread by mosquitoes and infects birds (especially the American crow), animals (horses), and people. West Nile disease caused an epidemic in 1999 in New York City, where it infected sixty-two people by year's end, claiming seven lives. The city moved quickly and launched an assault on mosquitoes with aerial pesticide missions and three hundred thousand cans of DEET. Though this kept the disease at bay for the next year, in 2002, West Nile disease spread with fervor and boasted 4,156 confirmed cases and 284 deaths and then spiked in the 2003 outbreak, during which 9,862 cases and 264 deaths were reported by the CDC. Since then the numbers have been toned down, though the disease has spread to every state and into Canada and Mexico. Like many chronic infectious diseases, its symptoms mimic other illnesses and often go undiagnosed and unreported. The infection is unnoticed by many, but some experience a stage of mild fever, similar to the common cold or flu, that lasts seven to ten days. From there, the disease can further invade the neurological system, causing severe meningitis and encephalitis, and in the immune-compromised it can lead to coma and death. There is no known cure, human vaccine, or herbal therapy proven effective; however, some of the invasive antivirals have strong potential. In his review of traditional Chinese medicine's approach to West Nile disease, Subhuti Dharmananda found herbs applied for the treatment of encephalitis and arboviruses (viruses that are transmitted by insects, which would include dengue fever, yellow fever, and West Nile virus). The herbs listed are readily available and have known broad-spectrum, antiviral effects, such as Japanese knotweed and Dyer's woad, also known as Ban lan gen (*Isatis tinctoria*), which is considered an invasive plant throughout the western U.S. rangelands. For the treatment of the encephalitis symptoms, specifically for reducing the fevers of infection, Dharmananda recommends berberine-rich medicinals, for which barberry could be substituted.[9]

Invasive Plants with Strong Potential for Treating West Nile Disease

- Japanese knotweed
- barberry
- dandelion
- sweet Annie
- Japanese honeysuckle
- purple loosestrife
- wormwood

Malaria is one of the most common infectious diseases throughout the tropical and subtropical world, and it was common throughout North America when European settlers first arrived here. Caused by a protozoan parasite of the *Plasmodium* species through the help of the pesky mosquito, it infects approximately three hundred fifty million to five hundred million people every year, killing between one and three million people, most of whom are children in sub-Saharan Africa. The symptoms of the disease are fever, chills, flulike symptoms, nausea, anemia-related illness (the parasites multiply in red blood cells), and, in severe cases, coma and death. We have been evolving the human genome and learning to survive with malaria for at least fifty thousand years, and plants were our first healers for relieving this sickness. Quinine and artemisinin (from the invasive plant sweet Annie, *Artemisia annua,* a relative of wormwood) are plant-based derivatives that are widely used as antimalarial drugs, and a wide array of other plants have been traditionally used and/or scientifically validated to address this widespread disease.

Invasive Plants Used to Treat Malaria

- wormwood   .
- English ivy
- tree-of-heaven

- sweet Annie
- plantain
- eucalyptus

Influenza

Influenza is one of the greatest scares we face on a yearly basis, and we go to great lengths to outsmart the upcoming seasonal virus and mutating forms of avian flu, swine flu, and so forth. In the winter of 2008–09, nearly all common types of flu were resistant to Tamiflu, and of growing concern were various strains that will mutate or have mutated. According to some scientists, if a flu pandemic of the same magnitude and severity as the one in 1918–19 were to occur in the present day, an estimated fifty-one to eighty-one million people worldwide would die. Yet people have coped with this disease for a long time, through healthy diets, supportive immune systems, and some weedy plants that grow all around us.

Invasive Plants Used to Treat the Flu

- Japanese knotweed
- multiflora rose
- wormwood
- sweet Annie

- Japanese honeysuckle
- barberry
- white mulberry
- mint

Tuberculosis

Tuberculosis (TB) is another disease that is growing resistant to the widespread use of antibiotics. The World Health Organization warns of killer strains that could run amok. *Mycobacterium tuberculosis* causes what is also known as consumption, a dreadful wasting disease that is passed on through germs floating through the air and is inhaled into the lungs. The infection can then spread to other systems of the body, including the brain and central nervous system and kidneys, creating an array of symptoms and leading to a slow death. According to WHO, tuberculosis is the top single infectious killer of adults worldwide, and it lies dormant in one in three people. Of those, 10 percent will develop active TB, and about two million people a year will die from it.

Even worse, the bacterium has developed multidrug resistance, rendering as most susceptible those in developing nations, those infected with HIV, and the elderly. Tuberculosis is still rare in the United States, but large-scale potential spread of this bacterium—with increased virulence— could result in devastation, as it did prior to the discovery of antibiotics.

Invasive Plants Used to Treat Tuberculosis

- artemisia
- Japanese honeysuckle
- sweet Annie
- thistle
- datura
- red clover

- barberry
- English ivy
- plantain
- Saint-John's-wort
- eucalyptus

Staph

Staphylococcus aureus (known commonly as the cause of staph infections) is one of the most widespread, resistant pathogens. It is naturally found in abundance throughout the mucous membranes and on many people's skin. Within four years of the mass production of penicillin, this bacterium built up resistance to it, and now at least half of all *Staphylococcus aureus* infections are resistant to the drugs penicillin, methicillin, tetracycline, and erythromycin. Vancomycin and linezolid were employed for this resistant bacterium in the late 1990s, and within a few years it began to grow resistant to these antibiotics as well. Methicillin-resistant *Staphylococcus aureus* (MRSA) and Community Acquired-MRSA are emerging epidemic threats responsible for the rapidly progressive fatal diseases that easily spread throughout hospitals and other areas of close human contact. It appears that MRSA is increasing in virulence with cross-contamination and transmission of the superbug from dogs and cats to humans, then back again, iultimately infecting homes, play areas, and beaches. In addition to stopping the use of pharmaceutical antibiotics to treat this deadly, resistant disease, using plant medicines will undoubtably slow its virulence.

Invasive Plants Used to Treat *Staphylococcus aureus*

- Japanese knotweed
- multiflora rose
- dandelion
- white mulberry
- bindweed
- Japanese honeysuckle
- tamarisk
- blackberry
- sweet Annie

Pneumonia

Streptococcus pneumonia is a pathogen responsible for pneumonia, bacteremia, otitis media, meningitis, sinusitis, peritonitis, and arthritis. It is increasingly resistant to penicillin, and using this and related antibiotics is implicated as a risk factor for acquiring a strep infection.

Invasive Plants Used to Treat *Streptococcus pneumonia*

- Japanese knotweed
- dandelion
- white mulberry
- Japanese honeysuckle
- common reed
- plantain

Waterborne Diseases

Researchers in preventive medicine are reporting on the dangers that extreme weather contribute to the spread of waterborne diseases by pathogenic bacteria, viruses, and parasites. Monsoonlike rainfall occurring in the cities and countryside increases the risk of disease outbreaks as storm water and wastewater combine and the resulting sludge is diverted into drinking water reserves and public swimming areas. The contamination of clean waterways leads to greater possibility of contracting waterborne infections such as *E. coli,* cholera, dysentery, salmonella, and giardia. In 1993, Milwaukee, Wisconsin, experienced an outbreak in city drinking water of the parasite *Cryptosporidium* to which were exposed more than four hundred thousand people, and more than fifty people died as a result. Outside urban areas, livestock sludge pools collect waste, which is then flushed downstream when heavy rains fall.

It's the perfect storm. Deteriorating urban water infrastructure, intensified livestock operations, and extreme climate change–related weather events may well put water quality, and thereby our health, at risk.

JONATHAN PATZ, UNIVERSITY OF WISCONSIN
PROFESSOR OF MEDICINE AND PUBLIC HEALTH[10]

Invasive Plants Used to Treat E. coli Infection

- Japanese knotweed
- multiflora rose
- white mulberry
- bindweed
- Japanese honeysuckle
- tamarisk
- wild mustard

Invasive Plants Used to Treat Cholera

- Japanese honeysuckle
- barberry
- blackberry
- Oriental bittersweet

Invasive Plants Used to Treat Dysentery

- barberry
- Japanese honeysuckle
- dandelion
- sweet Annie
- Siberian elm
- tamarisk
- purple loosestrife
- Oriental bittersweet
- common reed
- tree-of-heaven
- plantain
- English ivy

Invasive Plants Used to Treat Salmonella

- Japanese knotweed
- tamarisk
- barberry
- Japanese honeysuckle
- dandelion
- bindweed

Invasive Plants Used to Treat Giardia

- tree-of-heaven
- plantain
- Siberian elm
- barberry

The following chart sorts it in another way.

INVASIVE PLANTS AND THEIR
EFFECTIVENESS AGAINST INDIVIDUAL PATHOGENS*

PLANT	PATHOGEN
barberry	*E. coli, Salmonella typhi, Staphylococcus aureus, Streptococcus pneumonia, Mycobacterium tuberculosis* (TB), *Bacillus dysenteria, Vibrio cholera,* **Borrelia burgdorferi (Lyme disease),** *Candida,* cancer, **malaria,** *Giardia*
bindweed	*E. coli, Salmonella typhi, Pseudomonas aeruginosa, Staphylococcus aureus, Candida,* cancer
blackberry	*Staphylococcus aureus,* **Bacillus dysenteria,** *Vibrio cholera,* cancer
dandelion	*Salmonella typhi, Pseudomonas aeruginosa, Staphylococcus aureus, B-hemolytic streptococcus, Bacillus dysenteria,* **Borrelia burgdorferi (Lyme disease),** cancer
English ivy	**Mycobacterium tuberculosis (TB),** *Candida,* cancer
garlic mustard	cancer
Japanese honeysuckle	*E. coli, Salmonella typhi, Pseudomonas aeruginosa, Staphylococcus aureus, B-hemolytic streptococcus, Streptococcus pneumonia, Mycobacterium tuberculosis* (TB), *Bacillus dysenteria, Vibrio cholera,* **Borrelia burgdoferi (Lyme disease),** influenza, **cancer,** syphilis, **malaria**
Japanese knotweed	*E. coli, Salmonella typhi, Pseudomonas aeruginosa, Staphylococcus aureus, B-hemolytic streptococcus, Streptococcus pneumonia,* **Borrelia burgdofere (Lyme disease),** influenza, cancer
knapweeds	*Pseudomonas aeruginosa, Staphylococcus aureus,* **intestinal worms,** cancer, **measles (Rubeola), malaria**
kudzu	**Bacillus dysenteria,** cancer, measles (Rubeola)
Oriental bittersweet	*Pseudomonas aeruginosa, Staphylococcus aureus,* **Streptococcus pneumonia, Bacillus dysenteria, Borrelia burgdofere (Lyme disease), intestinal worms,** cancer, **syphilis**
plantain	*E. coli, Streptococcus pneumonia,* **Mycobacterium tuberculosis (TB),** *Bacillus dysenteria,* **Borrelia burgdofere (Lyme disease),** cancer, **syphilis,** malaria, *Giardia*

*Plain text = cited in scientific literature (in vitro or in vivo studies).
Bold = No scientific references; cited in traditional or modern uses or by author's analysis.

PLANT	PATHOGEN
purple loosestrife	*E. coli, Staphylococcus aureus,* **Streptococcus pneumonia,** *Bacillus dysenteria, Candida,* cancer
reed	**Pseudomonas aeruginosa,** *B-hemolytic streptococcus,* **Streptococcus pneumonia, Mycobacterium tuberculosis (TB),** *Bacillus dysenteria, Vibrio cholera,* cancer, measles (Rubeola), **Giardia**
Russian olive	*E. coli, Pseudomonas aeruginosa, Staphylococcus aureus,* cancer
Scotch broom	cancer
Siberian elm	**E. coli, Bacillus dysenteria,** intestinal worms, cancer, **Giardia**
sweet Annie	*Salmonella typhi, Pseudomonas aeruginosa, Staphylococcus aureus, Mycobacterium tuberculosis* (TB), *Bacillus dysenteria,* **Borrelia burgdoferi (Lyme disease),** influenza, intestinal worms, *Candida,* cancer, malaria, **Giardia**
tamarisk	*E. coli, Salmonella typhi, Staphylococcus aureus,* cancer, measles (Rubeola), **syphilis**
thistle	*Mycobacterium tuberculosis* (TB), cancer
tree-of-heaven	**Streptococcus pneumonia, Mycobacterium tuberculosis (TB),** *Bacillus dysenteria,* intestinal worms, *Candida,* cancer, malaria, *Giardia*
white mulberry	*E. coli, Pseudomonas aeruginosa, Staphylococcus aureus, Streptococcus pneumonia,* **Borrelia burgdoferi (Lyme disease),** influenza, cancer
wild mustard	*E. coli,* **Streptococcus pneumonia, Mycobacterium tuberculosis (TB), cancer**
wild rose	*E. coli, Pseudomonas aeruginosa, Staphylococcus aureus,* influenza, **cancer**
wormwood	*E. coli, Salmonella typhi, Staphylococcus aureus,* intestinal worms, *Candida,* **cancer**

As they fell from heaven, the plants said,
"Whichever living soul we pervade, that man will suffer no
harm."

THE RIG VEDA

HEAVY-METAL TOXICITY

As we saw earlier in text addressing environmental concerns and phytoremediation, heavy metals and toxic pollutants that are accumulating in and around our bodies, homes, and ecosystems are another invading pandemic. They threaten our community's health, and each of us is affected. Used every day in agricultural, industrial, pharmaceutical, and household products are at least thirty-five metals of concern, and twenty-three of these are classified as heavy metals due to a specific gravity that is more than five times the specific gravity of water. Heavy metals are contained in pesticides, chemical fertilizers, paints, cosmetics, cookware, computers, and medicines and in almost all human-manufactured items. We make use of these compounds on a daily basis. Sometimes without our knowing it, professions and hobbies bring us into intimate contact with these toxic elements that slowly leach into the body. Some of the workers who may come in contact with these metals: physicians, pharmaceutical manufacturers, laboratory workers, dentists, hairdressers, painters, welders, landscapers, farmers, photographers, and potters.

After the metals are released into the environment (water, soil, air), they are absorbed into our bodies through inhaling, ingesting, or contact with the skin. Increasingly, we hear warnings of toxic contamination that can affect our well-being. Indeed, some of us are beginning to realize that serious health issues can arise due to chronic exposure to these poisons. Many of these elements exist in a stable metallic form, so they are not absorbed or reconfigured but instead accumulate in the food chain and eventually in the body. Heavy metals have no known utility in the body, and they (mercury, lead, cadmium, aluminum) are toxic in minute amounts. Other minerals (chromium, copper, molybdenum, nickel, selenium, zinc, and possibly arsenic) in ionic form can be assimilated and are needed by the body in small amounts, but even excesses of these essential metals can collect in tissues, where they cannot be metabolized or broken down efficiently or they can inhibit other minerals from being processed and used by cells. For example, when lead builds up in the body, it replaces the calcium in the bones and is stored there. It then hinders the other basic functions of calcium in the human body and can result in osteoporosis due to insufficient calcium. Similarly, excess zinc in the system can lead to deficient copper in the body, and the reverse is aso true: excess copper depletes zinc stores in the body. Any excess of these metals slowly accumulates in the body and inhibits basic functions of various bodily systems, and there are few methods to assess this buildup. Blood tests can reveal the presence of acute and excessive accumulations of heavy metals in the body, and hair analysis is another popular means to diagnose metal toxicity deep within the tissues and bones.

Over time, as the heavy metals displace the essential minerals in the body, people experience greater health problems that are vague in nature and difficult to track. In general, toxic metal buildup leads to free radical proliferation, alters blood cell development, can accumulate into tumor growth, can damage central nervous functioning, and can influence all vital organ systems—over time, deeply impacting a person's overall physical, mental, emotional, and energetic health. Some of the degeneration caused by this toxicity can result in central nervous system disorders that lead to a variety of complaints. These are often misdiagnosed as other diseases, such as Alzheimer's disease, Parkinson's disease, muscular dystrophy, Lyme borreliosis, and multiple sclerosis.

Other types of chemical toxicity alter hormone levels, creating distortions and requiring great adjustments of the lymphatic system in order to balance the functioning of the adrenal, thyroid, and pituitary glands. Many epidemiological studies show that increased exposure to toxic chemicals leads to a greater chance of experiencing various cancers. Non-Hodgkin's lymphoma, for example, is the most rapidly increasing malignant disease. Some of the chemicals suspected to contribute to its rise include: phenoxyacetic acids, chlorophenols, dioxins, organic solvents (including benzene), polycholorinated biphenyls, chlordanes, and immunosuppressive drugs. The effects of these toxic substances, as well as heavy metals, have become readily apparent in some, while others suffer low-grade chronic illness without recognizing the source because this controversial affliction is difficult to diagnose. People continue to be poisoned, though mostly unaware due to the saturation of each breathing moment of our lives by toxic chemicals and heavy metals.

INVASIVE HEAVY METALS

Lead (Pb) is found in some fertilizers and therefore on some foods, especially fruits and grain. It was unregulated until 1978, so old and peeling paint, sodder in pipes that carry drinking water, and the toys in Grandma's attic can all contain substantial amounts of lead. These days, almost everything made in China contains lead. Because regulation is nonexistent and children are the most susceptible to lead toxicity due to their growing bodies and their general interest in putting everything in their mouths, lead poisoning has become an alarming concern for many families. According to the National Institutes of Health (NIH), lead poisoning in children can lead to "reduced IQ, slowed body growth, hearing problems, behavior or attention problems, failure at school."[11]

Aren't these symptoms that doctors prescribe Ritalin for?

Some of the symptoms of lead poisoning may include "irritability, aggressive behavior, low appetite and energy, difficulty sleeping, headaches, reduced sensations, loss of previous developmental skills (in young children), anemia, constipation, kidney damage, abdominal pain and cramping (usually the first sign of a high, toxic dose of lead poison), and very high levels may cause vomiting, staggering gait, muscle weakness, seizures, or coma."[12] Long-term studies are beginning to show chronic exposure can lead to violent crime and shrinkage of the brain's gray matter (influencing intellect, reasoning, empathy) and will also affect the reproductive system, strength of bones and joints, and blood cell synthesis.[13]

Mercury (Hg) is a notorious and controversial metal, and even though the dangers of mercury poisoning have long been known, this element continues to expand its scope of influence. Since the 1930s, mercury has been used as a preservative in the vaccines and medicines routinely injected into people's bodies and has been increasingly found in the fish we eat. It is in drinking water and leaches from dental amalgam through reactions with the acidic foods people eat. According to the EPA, the most prevalent sources of mercury come from power plants, cement kilns, refineries, and commercial boilers, with an estimate of more than 112 tons of mercury emissions generated in 2005. In 2001, the CDC reported that one in ten American women who have babies are at risk of in utero mercury exposure, which seems to "permanently damage the brain, kidneys, and developing fetuses," according to the federal Agency for Toxic Substances and Disease Registry.[14]

Arsenic (As) is a common component in fertilizers and pesticides and is most often leached into conventional root vegetables such as onions, carrots, and potatoes. This highly toxic and carcinogenic substance is also found in pressure-treated wood, and from there it leaches into the surrounding drinking water. For centuries it has been explored as a medicine, as an antibiotic, as useful in cancer therapy, and as a radioactive isotope in PET scans. Arsenic is carcinogenic, disrupts basic cellular functioning (hence some of its medicinal uses), and will lead to multisystem organ failure and death in excessive poisoning. Even with the strong toxicities this element imparts, micronutrient forms of arsenic might be a trace essential mineral for the body. Evidence is mounting in the area of amino acid metabolism that shows that arsenic appears to be needed for proper metabolism of both methionine and arginine.

The extensive use of *Copper* (Cu) as a conductor has led to vast applications that, over the years, have turned into large wastes in highly concentrated forms. Copper pipes could pose problems if they carry highly acidic water, and copper cookware reacts to acidic foods, leading to cirrhosis of the liver. Also used in agricultural pesticides, it can contaminate food and affect estrogen levels. Copper toxicity can be exacerbated by insufficient zinc in our diet, which is common among vegetarians, highly stressed individuals, and denatured-food junkies who eat only refined and processed food. Symptoms of copper poisoning are similar and vague, like other metal toxicities, with incidences affecting the central nervous system and various organs and sometimes leading to liver disease.

Aluminum (Al) is found in cookware, foil, canned foods, over-the-counter medications, antiperspirants, and douches; produces a heavy pesticide residue on conventional produce; and is used as a food additive in products people often consume. Once again, this metal is a central nervous system disruptor, which then affects all aspects of the essential bodily functions, leading to mental and emotional issues, headaches, speech problems, anemia, digestive impairment, and kidney dysfunction. Aluminum also hinders the body's ability to digest and assimilate calcium, phosphorus, and fluoride, leading to osteoporosis and loss of bone growth and density.

Cadmium (Cd) is a lesser-known heavy metal in our environment and bloodstream. Though banned in the European Union, this metal is used in rechargeable batteries; as a plastic stabilizer in PVC; for brilliant blue color in paints; and, because of its photosensitivity, is used in televisions, solar panels, and photocopiers. It is also abundant in cigarette smoke. Cadmium and cadmium compounds have been spread throughout the environment. In the soil, it has a half-life of fifteen to one hundred years, most often impacting crops such as lettuce, corn, and wheat. Excessive exposure to cadmium may result in cancer, kidney disease, neurological damage, infertility, and birth defects.

INVASIVE PLANT CHELATION: PLANT REMEDIATION FOR THE TOXIC BODY

If heavy metals and pollutants are left to accumulate in the body, they become free radicals, altering chemical structures and injuring cellular DNA. They begin accelerating the process of oxidation by creating chain reactions of mutated cells that can accumulate and hinder basic bodily processes. Although oxidation is a natural aging process that happens to all living cells—as evidenced by the browning of an apple, composting food, the rusting of an iron pan, the healing of an exposed cut—it can become problematic when the body is overburdened and is not supplied with essential compounds to aid healing. Oxidation-formed free radicals are seeds for disease that lead to various symptoms and disorders. Plants provide antioxidant compounds as healing salves for wounded cells, greasing the rusty parts and helping to mitigate the effects of oxidation by removing the damaged cells and encouraging new growth. All of the time, these compounds are discovered and verified within various fruits, vegetables, grains, and nuts, contributing their cancer-protective, life-enhancing qualities. Antioxidants are found in abundance throughout the green kingdom in various fruits, vegetables, tea, *and* many invasive plants. These prized compounds—such as vitamin C, beta-carotene, rutin, resveratrol, luteolin, catechin, kaempferol, and quercetin—have been found to be powerful healing agents that scavenge for free radicals, protect against cancer, slow the progression of age-related degeneration, and act as the basis of supportive chelation therapy.

Chelation therapy involves removing heavy metals from the tissues in the body using special agents or binding forces that help excrete the toxins. EDTA, a synthetic amino acid, is the chelating agent most widely used in therapy and was first developed for industrial lead toxicities. Antioxidants, chlorophyll-rich greens and seaweeds, liver detoxification herbs, garlic, bentonite clay, trace minerals, and homeopathic sulfur have all been employed in natural detoxification regimes, and each seems to amplify the overall effects of EDTA. Cilantro is the notable herb to address heavy-metal toxicity, helping to remove mercury, lead, aluminum, and possibly other heavy metals from the central nervous system and bodily tissues. Often used are the bitter, liver-supportive herbs such as dandelion, burdock, yellow dock, and milk thistle that help aid detoxification and prevent toxic liver damage. The use of trace minerals to bind the excess toxic metals in the body appears to help, and because some toxicities can inhibit the absorption of other precious trace elements, our bodies can benefit from mineral supplementation that can counter the side effects of the toxicity. Science is exploring other plants and phytochemicals that can aid the body in this way, and because many of these plants have direct connections to removing toxic metals and chemicals from the environment, we can therefore see a potential for the human ecology as well.

We can phytomine the toxic human landscape.

As discussed throughout chapter 7, addressing phytoremediation, plants have the capacity to absorb and nullify toxic substances within the soil, impeding any further penetration of these substances into the surrounding environment. Many plants possess unique phytochelating compounds that can deal with heavy metals and pollutants by attracting, stabilizing, and removing them from the land, water, and air. Plants stop the proliferation of these spoilers of the soil and the free radicals oxidizing the environment by acting as antioxidants for the earth's biosystems. Most of these plants possess their own *phytochelators* to help process these toxic substances that stress them and impact their growth. Metal-chelating plant compounds (metallothioneins, phytochelatins, phytins, organic acids) have similar roles—and are often the same compound—as the flavonoids, amino acids, enzymes, vitamins, and minerals that have been explored for their use in chelating the human body. According to noted phytotherapist David Hoffmann in *Medical Herbalism:* "Many flavonoids have the ability to bind heavy metal ions that are known to catalyze the production of free radicals . . . [and] from a traditional herbal perspective, many links can be made between research-based reports of flavonoid activity and the . . . actions of plants in our materia medica."[15]

According to researchers in the field of pharmacognosy, "the flavonoids [contained in many invasive plant species] have certain health effects, and their antioxidant, radical scavenging, anti-mutagenic and anti-carcinogenic properties are well known."[16] Upon review of the compounds contained within many plants, the evidence begins to reveal expansive antioxidant activity that can address different kinds of heavy-metal toxicity. There is very little research concerning the use of herbal medicines in the treatment of illness due to toxic contaminants, though one scientific study reveals the detoxification ability of Japanese knotweed. This prolific plant is known to gravitate to contaminated environs, is a rich source of medicine containing potent antioxidants and trace phytonutrients, and has been used successfully in the treatment of leucopenia due to exposure to toxic chemicals used in radiation and chemotherapy and from exposure to naphthalene and benzene (ubiquitous toxic chemicals derived from coal tar). Opportunistic plants, in general, appear to meet the criteria for further study by possessing numerous antioxidant compounds and because many of them have been used traditionally as detoxifying, regenerating, and immune-supporting agents for healing.

In cases of chronic exposure to toxic metals, the comprehensive detoxification process involves addressing multiple layers in order to treat not only the metal toxicity but also the person afflicted with careful consideration of his or her constitution, immune health, and any other underlying disease. Specific binding agents are known to adhere to different metals (just as in phytoremediation projects) and are combined with antioxidants, which, by definition, will prevent free radical oxidative damage that takes place in chronic toxic exposure. In conjunction with chelating the metal toxicity, providing general support to the organs of elimination helps the detoxification process. Supporting healthy liver function is emphasized to clean the blood, assist basic metabolism, and aid the elimination of toxins. In addition to any malabsorption issues due to excess metal accumulation, some essential vitamins and minerals may begin to be leached out of the system through the detoxification treatment. Therefore, supplementation is necessary. Fortunately, many opportunistic plants and a wide array of fresh fruits, vegetables, and foods provide a balanced intake of vitamins and minerals, although, at times, potentized supplements can expedite the process of absorption. Most invasive plants possess these health-promoting components and also contain other compounds that can treat underlying diseases, disorders, or infections while they support the immune system and constitutional health of an individual.

Antioxidants and Chelating Compounds in Invasive Plants

In the weedy invasives that grow all around us there are an abundance of potent compounds for health, longevity, and curing diseases. Most of these plants contain different combinations of flavonoids, amino acids, vitamins, and minerals that have antioxidant, antimicrobial, and detoxifying effects to support the body in ridding itself of toxic accumulations. By combining these micronutrients as nature intended within an entire plant form, the human body can better assimilate them. Studies during the 1990s showed remarkable capabilities of these compounds, with the introduction of vitamin C, vitamin E, and zinc. Yet some studies also showed some detrimental side effects to using potent, isolated components, which curtailed some of the momentum of antioxidants as the next cure-alls. During the same time, not a single antioxidant food study showed negative results or had to be canceled.[17] This proves the importance of using these medicines in combinations that the body can handle and that have a greater spectrum of influence on toxic elements. Because each individual component within a plant has relevance to the context of use as a protecting and detoxifying chelating agent, its sphere of influence multiplies, and it is able to address a wider variety of toxins and assist other antioxidants in the process of healing cells and tissue.

> *Acknowledgment of the role of an isolated constituent in health maintenance should emphasize the importance of the many herbal and dietary sources of the compound, rather than focus on "natural magic bullets" available only in dietary supplement form.*
>
> DAVID HOFFMANN, *MEDICAL HERBALISM*

As time goes on, the list of antioxidants continues to grow, and the compounds found in invasive plants have been isolated, studied, and then manufactured by pharmaceutical companies, and these have become profitable drugs that fill grocery and pharmacy shelves.

Vitamin C, or ascorbic acid, is an essential compound for human health and is found throughout the plant world. Vitamin C is a powerful antioxidant used for its immune-enhancing properties, its metal-chelating capability, and the protection it offers the body against the effects of stress.

Invasive Plants Containing Vitamin C (Ascorbic Acid)

- garlic mustard
- blackberry
- wormwood
- barberry
- plantain
- black mustard
- lamb's-quarters
- chickweed
- pigweed (*Amaranth*)
- yarrow
- Japanese knotweed
- wild rose
- Russian olive
- dandelion
- common reed
- Saint-John's-wort
- sea buckthorn
- pokeweed
- yellow dock
- mullein

Artemisia (*Artemisia absinthium*), also known as wormwood, full view

Artemisia (*Artemisia absinthium*), leaves

Barberry (*Berberis vulgaris*);
photo by Steven Foster

Blackberry (*Rubus discolor*)

English ivy (*Hedera helix*);
photo by Larry Scott

Garlic mustard (*Alliaria petiolata*);
photo by Steven Foster

Japanese honeysuckle
(*Lonicerae japonica*); photo by Steven Foster

Japanese knotweed
(*Polygonum cuspidatum*), full view

Japanese knotweed (*Polygonum cuspidatum*), leaves and flowers

Spotted knapweed (*Centaurea maculosa*)

Oriental bittersweet (*Celastrus orbiculatus*);
photo by Steven Foster

Plantain (*Plantago major*)

Purple loosestrife (*Lythrum salicaria*)

Common reed (*Phragmites australis*)

Autumn olive (*Elaeagnus umbellate*)

Scotch broom (*Cytisus scoparius*);
photo by Steven Foster

Siberian elm (*Ulmus pumila*), full view;
photo by Julie McIntyre

Siberian elm (*Ulmus pumila*), leaves;
photo by Julie McIntyre

Tamarisk (*Tamarisk*);
photo by Julie McIntyre

Bull thistle (*Cirsium vulgare*)

Canada thistle (*Cirsium arvense*);
photo by Steven Foster

Tree-of-heaven (*Ailanthus altissima*);
photo by Julie McIntyre

White mulberry (*Morus alba*);
photo by Steven Foster

Wild mustard (*Brassica rapa*)

Wild rose (*Rosa rugosa*)

Beta-carotene is the important precursor to vitamin A in the class of carotenoids and is found throughout the plant kingdom. Vitamin A is especially known for aiding the eyes and improving night vision, though it is also essential for growth and development, strengthening immunity, fighting infection, helping the healing process, and in many cases protecting against cancer. It is most notable in the brilliant yellow and orange colors of roots, flowers, fruit, and animals—living life-forms such as carrots, sweet potatoes, marigolds, daffodils, tomatoes, peppers, flamingos, and starfish (and people who drink too much carrot juice).

Invasive Plants Containing B-carotene Carotenoids

- wormwood
- garlic mustard
- milk thistle
- white mulberry
- kudzu
- blessed thistle
- pigweed
- mullein
- yellow dock
- sea buckthorn
- black mustard
- dandelion
- Russian olive
- dodder
- wild rose
- pokeweed
- lamb's-quarters
- mugwort
- nettle

Rutin is a plant constituent recognized for its antioxidant, anti-inflammatory, anticarcinogenic, and cytoprotective actions. It is often noted in buckwheat, black tea, and apple peels. It has been reported to scavenge free radicals and chelate iron to protect cells from damage.

Invasive Plants Containing Rutin

- common reed
- wormwood
- Japanese knotweed
- Japanese honeysuckle
- multiflora rose
- puncture vine
- coltsfoot
- barberry
- red star thistle
- English ivy
- white mulberry
- jimsonweed
- yarrow
- Saint-John's-wort

A study by UCLA cancer researchers found that people who ate a diet rich in the flavonoids *quercetin, kaempferol,* and *catechin* appeared to be protected from developing lung cancer.[18] An additional eight-year study found that quercetin, kaempferol, and myricetin reduced human pancreatic cancer risk by 23 percent.[19]

Quercetin is one of the most prolific flavonoid compounds in the natural world. It has been found to scavenge free radicals effectively and to protect from heavy metal toxicity and various cancers. As a potent antioxidant, quercetin chelates iron from the body and has been shown to protect against cadmium nephrotoxicity and arsenic-induced oxidative stress.

Catechin is a highly abundant compound in tea, chocolate, fruits, vegetables, wine, and some invasive plants (and the controversial poison emitted from the roots of knapweed). Catechin is used therapeutically for its antimicrobial, anti-inflammatory, and antihepatotoxic activity.

Invasive Plants Containing Quercetin

- blackberry
- tree-of-heaven
- barberry
- tamarisk
- Russian olive
- kudzu
- wild rose
- Saint-John's-wort
- goldenrod
- coltsfoot
- euphorbia
- Japanese honeysuckle
- Japanese knoweed
- star thistle
- Scotch broom
- wormwood
- white mulberry
- dodder
- milk thistle
- puncture vine
- yarrow
- sea buckthorn

Invasive Plants Containing Kaempferol

- blackberry
- barberry
- Japanese honeysuckle
- multiflora rose
- boneset
- Russian olive
- English ivy
- white mulberry
- dodder
- sea buckthorn

- privet
- milk thistle
- puncture vine
- poison ivy
- euphorbia

Invasive Plants Containing Catechin

- knapweed
- Japanese knotweed
- Oriental bittersweet
- Russian olive
- wild rose

Resveratrol is the antioxidant praised in the "French Paradox" of rich and creamy foods and all-important red wine. It is derived from red grape skins (which contain many other antioxidants). Resveratrol is actually a medicine for the plants that possess it, a type of polyphenol called a *phytoalexin,* which is a compound used as part of the plant's immune system against invading fungus, disease, stress, injury, infection, and ultraviolet irradiation. Resveratrol has been studied extensively and heavily patented and has shown effectiveness against numerous cancers, tumors, infections, and inflammations. It has wonder-drug effects and antiaging attributes. Japanese knotweed contains more resveratrol than any other plant on Earth, and white mulberry is another invasive plant with this powerful component.

Invasive Plants Containing Resveratrol

- Japanese knotweed
- white mulberry

Luteolin, a powerful antioxidant, is believed to reduce inflammation, promote carbohydrate metabolism, benefit the immune system, scavenge free radicals, and prevent against cancer.

Invasive Plants Containing Luteolin

- Scotch broom
- common reed
- Japanese honeysuckle
- dandelion
- plantain
- blessed thistle
- mint
- yarrow
- bishop's weed

Chlorogenic acid, an ester of caffeic acid and quinic acid, is said to be a powerful antioxidant with cancer prevention properties that is able to help slow the release of glucose into the bloodstream after a meal. It is a wide-spectrum antimicrobial and has been used as an anti-infection ingredient in pharmaceuticals, cosmetics, and food. Chlorogenic acid has relatively low toxicity and few side effects.

Invasive Plants Containing Chlorogenic Acid

- purple loosestrife
- black mustard
- Russian olive
- Japanese honeysuckle
- plantain
- jimsonweed
- red clover
- European goldenrod
- blackberry
- English ivy
- wormwood
- white mulberry
- southernwood
- burdock
- chamomile

Caffeic acid is in the class of cinnamic acids and is found in abundance throughout the fruit and vegetable (and coffee) world. It is another potent compound with analgesic, antimicrobial, anticarcinogenic, anti-inflammatory, and immunomodulatory actions. Caffeic acid and the derivative caffeic acid phenethyl ester (CAPE) have been explored and found to be antioxidants with potent healing virtues. Caffeic acids are contained in many plants and trees and in propolis made by honeybees. These compounds have been shown to protect against liver and skin cancers, reduce inflammation, and aid the immune system with toxic elements.

Invasive Plants Containing Caffeic Acid

- Japanese honeysuckle
- barberry
- Russain olive
- plantain
- dandelion
- artemisias
- black mustard
- white mulberry
- common reed
- Canada thistle

- burdock
- yarrow
- jimsonweed
- nettle

- chicory
- chamomile
- privet
- red clover

Oleanolic acid is another antioxidant found throughout the plant kingdom that has been found to exhibit antitumor, hepatoprotective, and antiviral properties—including acting against HIV—and is said to be relatively nontoxic.

Invasive Plants Containing Oleanolic Acid

- Japanese honeysuckle
- Plantain
- blessed thistle
- privet
- self-heal

- Russian olive
- English ivy
- lamb's-quarters
- pokeweed

Sulfur (S) has been used for hundreds of years in homeopathy, Epsom salts, and hot springs to remove toxic buildup in the tissues of the body. Garlic is widely honored in chelation therapy, and this is due to to the fact that it contains at least thirty-three sulfur compounds and seventeen amino acids, most notably allicin, the compound that also gives garlic its smell and antibiotic properties.

Invasive Plants Containing Sulfur and Sulfur Compounds

- garlic mustard
- black mustard
- plantain
- tamarisk
- chickweed

- dandelion
- Japanese knotweed
- white mulberry
- nettle

Selenium (Se) and *Zinc* (Zn) are essential trace elements that we need for a healthy central nervous system and optimal immune functioning. Zinc is involved in more than one hundred enzyme reactions. These two act as antioxidants to protect against free-radical damage and to support the elimination of chemotherapy drugs while minimizing their side effects. Unbalanced levels of selenium and zinc also appear to cause damage to the brain and central nervous system. Increased rates of brain tumors co-occur with low levels of selenium, and often low levels of zinc have been found in cases of Alzheimer's disease. In addition, when zinc amounts tend to be insufficient, copper appears to accumulate, leading potentially to chronic toxicity.

Invasive Plants Containing Selenium

- barberry
- blessed thistle
- comfrey
- chickweed
- quackgrass
- dandelion
- milk thistle
- mullein
- nettle

Invasive Plants Containing Zinc

- barberry
- Japanese knotweed
- milk thistle
- Japanese honeysuckle
- Russian olive
- burdock
- nettle
- self-heal
- dandelion
- black mustard
- kudzu
- white mulberry
- comfrey
- privet
- coltsfoot
- chickweed

Molybdenum (Mo) is a trace element found in minute concentrations within the body. It is in numerous enzymes and influences protein synthesis, metabolic functions, and growth. It also has the ability to bind to copper and thereby help excrete it through the urine—and in extreme situations, to create a copper deficiency.

Invasive Plants Containing Molybdenum

- Japanese knotweed
- nettle
- dandelion

Bear in mind that yet again we have a situation where a chemical used in isolation will only give of its best when used as nature provides it—in combination and with the synergy of the whole plant.

DAVID HOFFMANN, *THE HERBAL HANDBOOK*

SUPPORT OF THE BODY'S FILTRATION SYSTEMS

The earth's intelligence in remediating toxic elements from the environment is reflected in the microcosm of the ingenious detoxification process of the human body. The liver, lungs, stomach, intestines, kidneys, bladder, and the lymph system all provide essential means to process daily environmental influences. In classical herbal traditions, plants possess a natural affinity with certain organs and organ systems and have been noted to help with the proper administration of medicine to afflicted areas. Many of these invasive plants used for their antimicrobial actions and for the detoxification of heavy metals have a specific impact on one (or more) of the filtering organ systems, and the direction of the herb can help facilitate the elimination of the toxin. We must note, though, that even with regard to the seemingly simple connection of a plant to an organ system, whole plant medicines work in a multitude of ways within the entire human organism, addressing different areas all at once. Each and every plant has unique, identifying characteristics, and analyzing subtle differences among the actions of plants brings about greater efficiency in healing unique individuals. In a similar way, plants are adapted to unique environments and interact with unique elements.

Just as they serve as filters of wastewater, contaminated landscapes, and disturbed ecosystems, invasive plants serve the human filtration system, helping the body ecology cope with the same toxic elements. During any detoxification process, one should give special attention to promoting healthy function of the liver and the detoxification of the blood. The liver filters the blood; breaks down complex proteins, fats, and carbohydrates in order to nourish the blood; stores toxic wastes brought into it from the tissues; and then releases them through the bloodstream into the excretory systems. Accordingly, plants with a bitter flavor stimulate the liver and gall-bladder function, enhance digestive function, and help aid in the circulatory and excretory systems. Herbs defined by their hepatic and alterative actions influence the liver and blood. Some weedy liver herbs include:

- dandelion
- barberry
- wormwood
- tamarisk
- milk thistle and thistle species in general
- plantain
- white mulberry
- burdock
- Japanese honeysuckle
- Oriental bittersweet
- Scotch broom
- Canada thistle
- knapweed (*Centaurea* spp.)
- Russian olive
- nettles
- yellow dock

The lungs are the air vents of our internal environments. They bring in fresh oxygen to supply the cells of the body, and in turn they release carbon dioxide and other gaseous wastes from the body. The deep rhythm of breathing aids in the circulation of oxygen-rich blood to all tissues and organs, providing the essential component to life on all levels. The skin is seen as a kind of third lung, also aiding in respiration and the elimination of toxins. Skin eruptions, blemishes, and rashes can all be indicative of an overloaded and toxic respiratory system. The diaphoretic and alterative actions in herbs help address eruptions of the skin, and expectorant, pulmonary, and anticatarrhal compounds are specific to the lungs. Some invasives that have influence on the lungs include:

- tree-of-heaven
- Japanese knotweed
- Siberian elm
- white mulberry
- common reed
- Japanese honeysuckle
- plantain
- wild mustard

The intestines are the body's sewage system, collecting its sludge, extracting the water and nutrients, and then eliminating that which the body does not need. When sewers back up, we know what happens. When the bowels do not work properly and there are excess toxins in the water and waste, the toxic sludge accumulates and can be reabsorbed into the bloodstream and lead to further irritations within the system. The therapy of purging is well known and rarely pursued, and it has been losing steam since the Middle Ages, though assistance is essential when the intestines are burdened with toxins and parasites. Plants capable of gastrointestinal detoxification generally are considered laxative and purging by nature:

- bindweed
- tree-of-heaven
- Siberian elm
- wormwood
- purple loosestrife
- dodder
- euphorbia
- common reed
- Japanese honeysuckle
- plantain
- kudzu
- tamarisk
- wild rose

The kidneys are responsible for filtering the water wastes of the body into the bladder for elimination of the acidic metabolites that build up in the blood. The kidneys are linked closely to the adrenal system, and when either system is burdened or weakened, the other feels the repercussions. The adrenal system is tied to the basic metabolism of the body and its innate immune response. When the adrenals become fatigued through stress, chronic illness, overwork, excess coffee, and so forth, the kidneys' processing of toxic metabolites is impaired, which can lead to further kidney disease and malfunction. Herbs with a diuretic function affect the kidneys and urinary system, and many plants considered tonics, alteratives, and adaptogens provide systemic support and replenish the adrenal system. These include:

- plantain
- dodder
- Siberian elm
- wild rose
- dandelion
- common reed
- Oriental bittersweet

As we have seen, many invasive plants are utilized as detoxifiers, antimicrobials, and immune stimulants, possessing countless vitamins, minerals, antioxidants, micronutrients, and other unknown potent phytochemicals. These attributes lend to their efficacy in chelation therapy and provide further assistance to the body for its elimination of environmental toxins. Responding to toxic increases in the environment, plants create these compounds for their own use and protection, to clean toxic metal accumulations from their environment. Because we humans share habitats and environs with local plants, however, their stressors are our stressors, and it follows that their coping strategies and healing virtues are our health solutions—as they have been for millennia. The potent compounds plants possess blended within the synergistic effects of the whole plant are proving to have broad-spectrum use against various emerging health threats—whether bacterial, viral, parasitic, or toxic. In the final section of this book, the individual plants will speak for themselves in their own voices to share their inherent wisdom and healing secrets.

> *Invasive plants are the warriors of the landscape, pioneering toxic and infected terrains, where they fight in the battle against the pathogenic influences of our ecosystems. Sometimes they engage the warrior within us and impart potent messages to our being so that we may make use of their medicine.*

CONCLUSION

With analysis of individual plant components, knowledge from traditional and modern-day medicinal uses, and understanding from the greater context of the plant's ecological dynamics, we can discover the use of invasive plants to treat invasive pathogens and heavy-metal toxicity. From this multifaceted viewpoint, we see the plants as the messengers of the earth that reveal to us the dangers of silent toxins. These loud, rampantly spreading plants are the canary in the coal mine, and they are our healers in disguise. As a toxin or disease moves into an area, so does the plant that can treat that toxicity. Coming down to us from previous generations is an ancient way of revering plants as our relatives. This signals a clear and distinct symbiotic relationship that has existed throughout history. Many stories have been told of how plants selflessly serve and freely give of themselves to the will of humans—and in return, we have provided for plants and helped move them around. The plants all around us have innate intelligence and a way of being that relates to and intimately understands humankind as well as the microscopic world of microbes and toxins. With tireless healing service, plants aid the environment and assist its inhabitants in coping with the pressures of existence. Since time immemorial, natural gardeners and master herbalists the world over have known that plants show up where they are needed.

> *In this way, with this perspective, we humans may be grateful for the all-pervading presence of these plants in our landscape.*

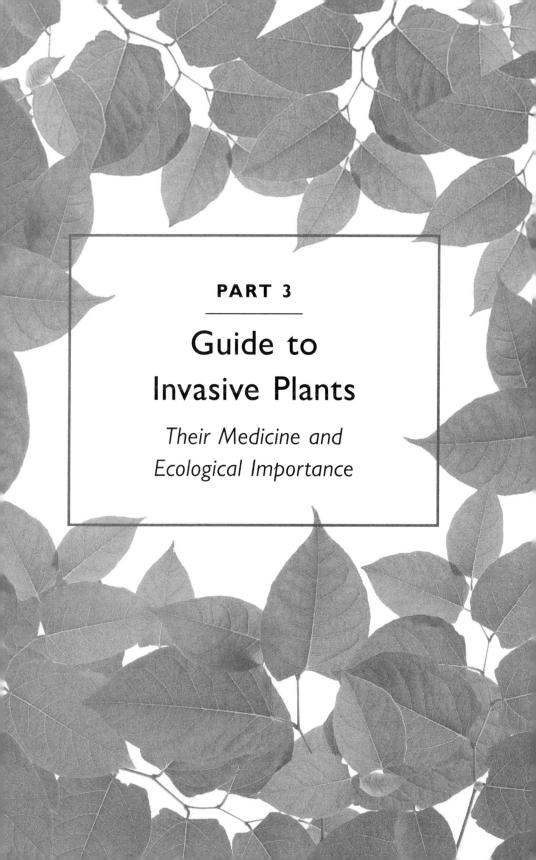

PART 3

Guide to
Invasive Plants

*Their Medicine and
Ecological Importance*

I HAVE CHOSEN THE following plants for their widespread appearance throughout North America. Some can be found in all areas, while a few are more isolated to regions but have a rampant presence there and have been found to be moving into other regions. Most all of the plants highlighted have come from Europe and Asia and have documented medicinal uses that span thousands of years. Many of these plants were brought to the United States by settlers for their medicine and benefits.

TERMINOLOGY AND CATEGORIZATIONS USED IN THIS MATERIA MEDICA

Below you'll find further clarification of some of the categories I use in the following guide.

- *Related Species:* Related species have a similar genetic lineage as the plant listed. This similar lineage means that these plant relatives have a genetic propensity for similar attributes and characteristics, such as similar compounds, growing conditions, genetic material, and so forth, allowing correlations to be made between them. Some related plant species have been tested for their medicinal effects and remediation potential in other areas of the world where they are used traditionally and not deemed "noxious."
- *Collection and habitat* refers to the areas where a plant lives and from where it came.
- *Parts used* refers to the plant parts used medicinally.
- *TCM* refers to traditional Chinese medicine. Included are the pinyin name of the herb and translation, taste, nature, and temperature, as well as the associated organ systems and channels according to TCM. This information is primarily from Bensky and Gamble's *Chinese Herbal Medicine Materia Medica,* and Chen and Chen's *Chinese Medical Herbology and Pharmacology.*
- *Ayurveda:* Much information was gathered from C. P. Khare's *Indian Herbal Remedies,* a thorough account of classical use, mod-

ern scientific insight, and crossover traditional uses of the ayurvedic Materia Medica.

- *Western Botanical:* Numerous sources were used for ancient and modern Western herbal knowledge, including two indispensible volumns of vast information: *A Modern Herbal* by M. Grieve and *King's American Dispensatory* by Felter and Lloyd.

- *Plant chemistry* refers to a list of known chemical constituents, with listings available in Dr. James Duke's *Handbook of Medicinal Herbs* and his web phytochemical and ethnobotanical databases available at www.ars-grin.gov/duke. Many of these compounds have potent influences on the body, but other constituents have no *known* healing qualities or human benefits. There are others still to be discovered.

- *Scientific studies:* Just because there are no studies detailing a plant's action does not mean the plant cannot perform this action. It is likely, instead, that there are yet no studies to confirm the specific action. This is most often due to lack of funding—the pharmaceuticals and plant compound extracts garner the most attention. Fortunately, many researchers in China and the Far East, India, and Germany are testing whole plant medicines that verify many of the traditional uses. The studies cited here were largely found through the National Institutes of Health database (www.pubmed.gov). The studies cited appear in each plant's References.

- *Harvesting and preparing:* The general harvesting tips given below apply to all plant species discussed. In cases that have special considerations in regard to harvesting or preparation, these appear under the plant itself. For more thorough information on medicinal harvesting and preparation, I recommend James Green's *The Herbal Medicine Maker's Handbook* and Richo Cech's *Making Plant Medicine,* or better yet, find a local herbalist who provides hands-on instruction.

- *Dosage:* Dosages are provided where possible, but due to varying methods of preparation, various conditions attributed to the herb, and variability among individuals, this section should be considered a very general, nonspecific guideline. See below for more on general dosage guidelines and the types of applications used (decoction, infusion, and tincture).

HARVESTING AND PREPARING

It is extremely important not to harvest plants intended as medicines or food from anywhere near or in contaminated soils or roadsides. This is especially true of most invasives because they tolerate heavy metals, clean the soil, and, in some cases, have been sprayed with herbicides even when they exist as established, healthy patches. Timing of harvesting varies based on whether you are harvesting the roots, leaves, or flowers. It is also important that when we harvest, we give thanks.

Roots
Roots should be harvested in the autumn, as the energy of the plant begins to return to earth after the warmer months of growth and bloom. The plant has stored up vital nutrients in its roots for winter hibernation, and these nutrients and compounds are sought after for medicine making. Of course, if the need of a plant root arises during another season, it is permissible to harvest.

Roots should be thoroughly washed of soil and debris, then chopped into smaller pieces and dried. They are dry when smaller pieces crisply snap. It is best to chop roots before they have dried, when they are still fresh and relatively soft. In most cases, once dried the root will become very difficult to cut. Once chopped, lay out the root pieces on a drying screen and provide good air circulation in a warm place out of direct light, though oftentimes, roots can be dried in the sun.

Leaves

In general, leaves are best picked fresh and in the springtime. This is when the plant's energy is returning upward in leaf growth. Most of the leafy greens we eat and make as medicine have cleansing and restorative powers for the blood and help the detoxifying ability of the liver. This is most important in the springtime, as we awake from the slumbers of winter and the richer foods of the season.

The leaves should be picked when clean and preferably in the morning. Wait until after the dew dries if dehydrating for storage. If the leaves are large, chop the leaves prior to drying, otherwise whole leaves will dry and shrivel, shrinking in size. If stems are harvested along with the leaves, cut them into one-inch pieces and dry until the stem snaps. Lay out leaves (and stems) on screens in a warm location with good airflow and out of direct sunlight.

Flowers

Flowers are most often associated with summertime, so this is the season to harvest. The flower is the fullest expression of a plant and its growth, imparting the spiritual color of the plant itself.

Flowers are picked fresh after opening, then processed by drying or are infused directly in oil, alcohol, or water. When drying, lay out the flowers one layer thick on a screen in a warm, dark place with good airflow. Because of the moisture content in most flowers, it is good to stir the flowers occasionally to prevent mold.

To store the dried herbs, place in cellophane bags, plastic bags, or sealed glass jars, in a cool, dry place away from direct light. Dried herbs will begin to lose potency after six months to one year. Tinctured herbs will stay potent for many years if kept in a dark bottle and in a cool location.

DOSAGE

Each herbalist has different preferences for doses and prescriptions, sometimes using multiple herbs, for most individuals they consult. Prescribing herbal dosages and combinations is as individual an artform as any, with a creative genius and invisible forces guiding the healer who works with plants.

Most of the dosing parameters come from TCM standards, in which the herbs are generally used with other medicinals in somewhat complex formulations. The dosages provided might prove as a reference to the strength of medicine, and you may want to compare this to other plants with similar strengths and known dosages.

The dosage subsection includes a reference to *grades* of dose.

high dose, food-grade medicine, which is very safe and to consume as desired
medium dose, strong-grade medicine, generally considered safe
low dose, very strong botanical, use caution and supervision

High dose implies that the medicine can be taken for longer periods of time, whereas the low-dose herbs should not be taken for any extended period of time.

General recommended herbal prescriptions dosages are:

3 or 4 times daily, with varying courses of days or weeks depending on the health situation. Consult a knowlegable herbalist or health care provider for more specific dosage information related to your health condition.

Types of herbal prescriptions:

Decoctions: The herbs with TCM dosages are generally prepared as a *decoction* in which they are cooked in two quarts of water for about twenty minutes to make a strong medicinal tea.

Infusions: The more delicate leaves and flowers may be prepared as an infusion, in which the herb is steeped in hot water for 10 to 12 minutes.

Tinctures: There are also some tincture dosages provided, in which the herb is soaked in an alcohol/water mixture for at least a month to extract the medicinal components from the plant.

Ointments and Poultices: To make an oil-based topical, infuse fresh or dried herb in oil (olive, sesame, or other oil or fat) and place in the sun or keep at a warm temperature (100 degrees F) for seven to ten days. For a poultice, make an infusion or decoction of the herb and place the plant material and tea-soaked cloth on the affected area of body. Liniments are alcohol extracts or tinctures that are used topically.

Artemisia
Artemisia absinthium var. *annua*

Common names: wormwood, absinthe, old man (*Artemisia absinthium*); sweet Annie, sweet wormwood, Oriental wormwood (*A. annua*)

Family: Aster family (Asteraceae)

Related species: The genus *Artemisia* consists of some three hundred species of annual, biennial, and perennial herbs and shrubs found throughout the dry, temperate regions of the world. Some of herbal interest include: mugwort (*A. vulgaris*), moxa (*A. moxa*), and tarragon (*A. dracunculus*).

States where wormwood or *Artemisia absinthium* is considered invasive

> *Earrings of the moon:*
> *Artemis, the goddess of wild things, chastity,*
> *   fertility,*
> *and the bloody hunt.*
>
> DALE PENDELL, *PHARMAKO/POEIA*

The Plant

Artemisia, in all forms and species, has gained notable attention throughout history by herbal physicians and medicine people. Traditional cultures separated by oceans and customs used artemisia to ward off pathogenic influences and evil spirits (which can be one and the same), and the purifying effects of the plant are well verified.

Medicines made with artemisia have often been said to be cure-all bitter potions. According to Stephen Buhner in *Herbal Antibiotics,* wormwood (*A. absinthium*) "is one of the most powerful herbs for the treatment of antibiotic-resistant disease available,"[1] and sweet Annie (*A. annua*) contains potent amounts of the isolated compound artemisinin, used to treat drug-resistant malaria.

I use A. annua *as a treatment for malaria and feel it to be one of the most effective treatments for it. When I travel in S. E. Asia with our groups I make everyone take it daily as a preventive for malaria.*

ROSEMARY GLADSTAR, THE GODMOTHER
OF AMERICAN HERBALISM

During the Vietnam War, China discovered the potent antimalarial effects of sweet Annie (*A. annua*), and provided to the North Vietnamese the medicine made from the plant. Artemisinin is now widely sold for treatment of malaria, and according to the World Health Organization, more than 1.5 million malaria patients in Southeast Asia and Latin America are treated with artemisinin every year. According to Stephen Buhner, however, this compound "produces a number of unpleasant side effects during treatment. However, when the whole herb is taken the side effects do not exist and treatment is just as effective. The plant has created several compounds whose only known purpose is to alleviate exactly the side effects that the isolated constituent produces in living organisms."[2]

Found in the essential oil of wormwood is thujone, a powerful compound that has strong effects on the central nervous system and that imparts the intoxicating properties of absinthe. The drink was befriended by many to alter consciousness, and it made poets of the crude and visionaries of the blind.

The Green Fairy emerges . . .
and you sing to the poet of your heart.
but watch out for that third glass.

The Ally
If you are sick, it will cure you.
If you are depressed, it will ease your soul—
If you are smitten, desirous to touch
the one so long beyond your grasp,
it will give you words
and a long green hour, together, to speak them.

DALE PENDELL,
PHARMAKO/POEIA

Description

Wormwood (*A. absinthe*) is recognizable by the silvery gray-green color of its feathery leaves. The leaves have a sagelike flavor and smell. It is a herbaceous perennial plant that grows 3 to 6 feet tall with stems arising from the base of the plant. In the summer, small yellow-brown flowers less that ¼ inch in diameter bloom along the tops of the stems and at the base of the leaves.

Sweet Annie (*A. annua*) is a bushy annual herb that grows from 1½ to 9 feet tall. The green leaves are sweet and pungent when crushed. They have toothed, lance-shaped individual leaflets with an overall soft, fernlike appearance that forms a broadly oval set of leaves 3 to 4 inches long. Small, round, and inconspicuous greenish yellow flowerheads (¹/₁₆ inch diameter) form in late summer to early fall.

Collection and Habitat

Wormwood is a native of Europe and long cherished by immigrants coming to the New World. It was imported and for sale in America by the early 1800s for gardens, as medicine, and as "flavoring" for drinks. Absinthe thrives in dry soils along roadsides, pastures, rangelands, and waste areas and is often found in these areas as large patches. It is now found throughout Canada and the United States but does not quite reach into the far southern states.

Sweet Annie is considered native to Eurasia, from southeast Europe to China, Russia, India, Korea, and Japan. It has now naturalized throughout North America and the world's waste places, disturbed lands, and roadsides.

Medicinal Uses

Parts used: leaf, flowering top, and essential oil

TCM: Qing hao (*A. annua*), flowering herb; bitter, acrid, cold; associated with the liver, gallbladder, and stomach

1. Clears summer heat.
2. Clears fever with accompanying weakness.
3. Cools the blood and stops bleeding.
4. Treats malarial disorders.
5. Clears liver heat and brightens the eyes.

Classically, *Artemesia annua* is used to treat summer heat with fever, headache, dizziness, stifling sensation in the chest, febrile diseases, rashes, nosebleed, red eyes, and light-sensitive eyes. By most regards, sweet Annie (*A. annua*) is the only *Artemisia* species that has appreciable amounts of artemisinin for treatment of malaria. TCM uses numerous *Artemisia* species with similar attributes, including *A. annua, A anomalae, A. apiacea, A. argyi, A. yinchenhao, A. capillaris,* and *A. scoparia.*

Ayurveda: Afsanteen (*A. absinthium*) (Arabic), flowering herb

Wormwood has been used classically in the treatment of liver and spleen inflammatory diseases, including hepatitis and enlargement of the organs and as a liver tonic. It has been prescribed for dysmenorrhea, fever, hysteria, epilepsy, nervous irritability, gastric and nervous depression, mental exhaustion, rheumatism, sprains, and bruises. *A. vulgaris* and *A. maritime* are also used for similar purposes.

Western botanical: According to Dr. John Hill (1772) in *A Modern Herbal,* "The leaves have been commonly used but the flower tops are the right part. These, made into a light infusion, strengthen digestion, correct acidities and supply the place of gall, where, as in many constitutions, that is deficient."[3] Wormwood is used as a bitter tonic with meals for those who have gallbladder disorders and to "supply the place of gall" and help digestive functions.

The Complete Commission E Monographs recognizes wormwood's efficacy in treating loss of appetite, dyspeptic complaints, and liver and gallbladder complaints.

With a wide range of antioxidants and antimicrobials and its effectiveness in stimulating the detoxifying functions of the liver, wormwood should be explored for chelation therapy. Wormwood contains thujone, antioxidants, rutin, vitamin C, beta-carotene, quercitin, and chlorogenic acid. In addition, it possesses strong tannins that have a potent effect on the brain and central nervous system, which therefore lends credence to the importance of wormwood in the removal of toxic overloads in the central nervous system.

Plant Chemistry

Artemisia absinthium: absinthin, achillicin, anabsin, anabsinin, anabsinthin, arabsin, artabasin, artabin, artabsonolide, artemetin, artemolins, artenisetin, arthamaridin arthamaridinine, arthamarin, arthamarinin, ascorbic acid, beta-carotene, B-caryophyllene, bisabolene, cadinene, camphene, chamazulene, chlorogenic acid, cis-epoxyocimene, eo, formic acid, hydroxypelenolide, inulosiose, isoabsinthin, isoquercitrin, isorhamnetin, ketopelenolides, lirioresinol, matrisin, nicotinic acid, p-coumaric acid, palmitic acid, patuletin, phellandrene, pinene, pepecolic acid, protocatechuic acid, quebrachitol, quercitin, rutin, sabinene, salicylic acid, spinacetin, syringic acid, tannin, thujone, thujyl-alcohol, vanillic acid

Artemisia annua: A-caryophyllene, A-pinene, arteannuin, artemisic acid, artemisia ketone, artemisilactone, artemisinin, artemisinol, artemisitine, artemisyl-acetate, B-selinene, B-sitosterol, borneol, camphene, camphor, capillene, caryohyllene-oxide, coumarin, cuminaldinene, deoxyartemisinin, eo, humulene, menthol, myrcenol, nonacosanol, o-cymene, ocimene, octocosanol, p-cymene, pinocarveol, qinghaosu-IV, quercetagetin-, scopoletin, stigmasterol, terpinen, tetratriacontane, ylangene

Pharmacological actions: Antimicrobial, digestive, anticholagogic, antitumor, anti-inflammatory

Scientific Studies

Antimalarial: The whole herb *A. annua* and the component artemisinin, which is most active in the leaves and flowers, have been proved to effectively kill malaria parasites in vitro and have become standard treatment and prevention in endemic areas.

Antimicrobial: The decoction of *A. annua* has been shown to inhibit in vitro *Staphylococcus aureus, Bacillus anthracis, Bacillus dysenteria, Pseudomonas aeruginosa, Neisseria catarrhalis, Corynebacterium diphtheriae, Mycobacterium tuberculosis, Candida albicans, Naegleria floweri, Klebsiella pneumoniae, Salmonella,* intestinal worms, and internal ameobic organisms. The artemisia essential oil from wormwood is effective against most microbes and has inhibitory effects on the growth of bacteria (*Escherichia coli, Staphylococcus aureus,* and *Staphylococcus epidermidis*), yeasts (*Candida albicans, Cryptococcus neoformans*), dermatophytes (*Trichophyton rubrum, Microsporum canis,* and *Microsporum gypseum*), *Fonsecaea pedrosoi,* and *Aspergillus niger.*

Anticancer: Artemisinin has shown inhibitory actions against various cancers both in vitro and in vivo. In vitro studies have found it effective against human leukemia cells, breast cancer, and colon cancer, while a study in dogs with bone cancer and cancer of the lymph nodes showed effectiveness with 10 to 15 days of treatment.

Antihypertensive: Injections of arteannuin into rabbit subjects lowered blood pressure, decreased heart rate, inhibited contraction of the cardiac muscle, and decreased blood perfusion to the coronary artery. In a similar study, it appeared to treat artificially induced arrhythmias.

Antipyretic: The herbal injection of *A. annua* was effective at reducing fevers in an in vivo study.

Cholagogic: The herb *A. annua* has been shown to increase the production and excretion of bile in rats.

Other Uses

Wormwood is used to produce the liqueur absinthe and is a key ingredient in vermouth or, as it is also known, Muse Verte (the Green Muse). Various *Artemisia* species have been used by numerous cultures to clear out pathogenic influences, both seen and invisible. The plants have been burned as incense or smudge to ward off insects, snakes, and evil spirits. Wormwood has also been used as an insect repellent by laying it among wool and fur to protect them from moths and insects.

Ecological Importance

In China, *Artemisia roxburghiana* was studied for its restoration potential with regard to an area contaminated with lead, zinc, and copper mine tailings. Of ten species, *A. roxburghiana* was found to be the most effective in reducing the contamination in the soil and was determined to be quite suitable for phytostabilization programs. A native of Canada, *A. frigid,* was found to be a cold-tolerant plant that could survive in crude oil–contaminated soil. With the ability to withstand such pressures without, in the process, losing biomass, this artemisia species is considered a good candidate for phytoremediation efforts. Western sage (*A. gnaphalodes*) was also found to be growing in a contaminated antitank firing range and was accumulating the explosive HMX from the soil.

Harvesting and Preparing

Research suggests that the artemisinin content of sweet Annie is highest from late July to August, just before flowering. Harvest the aerial parts of *Artemisia* on a dry day. Cut off the upper part of the leafy stems and tie into loose bundles. Hang the bundles out of direct light in a warm area and arrange the stems and bundles to allow enough air circulation to dry.

In ayurvedic medicine, wormwood is steeped in hot vinegar and applied externally for rheumatic inflammations, sprains, or bruises.

Dosage

Medium- to low-dose botanical. Often high doses are used for short periods of time.

Decoction: TCM, 3 to 9 grams (up to 30 grams). Should not be cooked for long as the active constituents of *A. annua* can be destroyed by excess heat, therefore some juice the fresh herb as a more effective treatment for malaria. A traditional prescription for using *A. annua* in malaria treatment calls for 30 grams decocted in 1 liter of water taken as one dose per day or 3 grams of fresh juice taken once a day.

Cautions and contraindications: Contraindicated during pregnancy. Use with caution in individuals with weak digestion.

Herb and drug interactions: It has been suggested that azole antifungals and calcium channel blockers present clinically significant herb-drug interaction with *A. annua* by inhibiting the formation of the artemisinin analogue, artelinic acid. Examples of antifungals include griseofulvin, nystatin, ketoconazole, fluconazole, and itraconazole. Examples of calcium channel blockers include nifedipine, amlodipine, diltiazem, and verapamil.

References

C. A. Groom, A. Halasz, L. Paquet, et al., "Accumulation of HMX (octahydro-1,3,5,7-tetranitri-1,3,5,7-tetrazocine) in Indigenous and Agricultural Plants Grown in HMX-contaminated Anti-tank Firing-range Soil," *Environmental Science and Technology* 36, no. 1 (January 2002): 112–18.

D. Lei and C. Duan, "Restoration Potential of Pioneer Plants Growing on Lead-Zinc Mine Tailings in Lanping, Southwest China," *Journal of Environmental Sciences* 20, no. 10 (China, 2008): 1202–9.

D. B. Robson, J. D. Knight, R. E. Farrell, et al., "Ability of Cold-tolerant Plants to Grow in Hydrocarbon-contaminated Soil," *International Journal of Phytoremediation* 5, no. 2 (2003): 105–23.

D. Lopes-Lutz, D. S. Alviano, C. S. Alviano, et al., "Screening of Chemical Composition, Antimicrobial and Antioxidant Activities of Artemisia Essential Oils," *Phytochemistry* 69, no. 8 (May 2008): 1732–38.

States where Japanese
barberry is considered
invasive

Barberry
Berberis vulgaris var. *thunbergii*

Common names: common barberry,
European barberry (*B. vulgaris*), Japanese barberry (*B. thunbergii*)

Family: Barberry family (Berberidaceae)

Related species: Oregon grape (*Mahonia aquifolium* var. *repens*), North American barberry (*B. canadensis*), Indian barberry (*B. aristata, B. asiatica*), *B. lycium*

The Plant

Barberry (*Berberis vulgaris*) sweeps through and takes hold in overthinned forests. With entwining barbed branches that spread along the forest floor, its establishment seems insurmountable. It grabs and scratches at us with each step we take, as if saying, "Stay out and don't come back!"

In addition to protecting and enriching the forest soils that have been disturbed, barberry has important medicinal uses for humans. This plant contains the rich yellow mark of berberine, the alkaloid of those long-honored herbal antibiotics known as goldenseal, Oregon grape root, goldthread, and Chinese coptis. Berberine was isolated as an herbal drug some fifty years ago in China, and it has been found to be a powerful constituent against bacterial, viral, fungal, parasitic, and yeast infections. This plant appears to be encroaching on humans in negative ways, yet barberry provides powerful healing compounds as it follows the trail of invasive epidemics (most notably, that of Lyme disease) and serves as a remedy for us (just as Dr. Qingcai Zhang of New York has used berberine-rich coptis for treatment of Lyme disease). In addition to berberine, barberry has antioxidant, detoxifying, and immune-supportive constituents for treatment of the presence of toxic substances. The plant contains important vitamins, minerals, and compounds, and a substantial proportion of these have individually proved effective as chelating agents. Barberry contains vitamins A, B, and C, selenium, zinc, quercetin, rutin, caffeic acid, kaempferol, and chlorogenic acid. In addition, it contains berbamine, an isolated component used for the treatment of depressed white blood cell counts due to the toxic overload of chemotherapy and radiation. Barberry is included as an ingredient in cancer treatment formulas (Hoxsey Cancer Formula), and berberine has been shown to protect the liver from toxic buildup of pharmaceuticals. In addition, a decoction of the bark is often prescribed for addicts to aid with withdrawal from opium or morphine. Where this invasive plant exists, barberry provides a gift to the human community as a potent plant medicine to help people cope with the surrounding pathogenic and toxic influences in their lives.

> *What I like about this plant is its ability to cling to rugged spaces. I've found it growing along the shale banks of Lake Champlain, hanging on with the old cedars for dear life, but quite rooted in and stable, in fact. It's this ruggedness I like in its medicine. A berberine-rich plant, it's often used as an analogue for goldenseal, but in my mind, they differ substantially. Barberry is very astringent, drying, and is even better than goldenseal for mucous inflammation and for mouth infections.*
>
> ROSEMARY GLADSTAR

Description

Barberry grows as a small shrub (2–3 ft.) to a large bush (up to 6 ft.) and is most notable in the autumn when the bright red berries load the branches. In the springtime, six-petaled flowers bloom along the barbed branches in clusters of two to four or singly. These form dry, mealy fruits that mature in the fall and can hang on throughout the winter to feed numerous birds. The oblong leaves ($\frac{1}{2}$ to 1 in.) vary in color from dark green to red, depending on the variety and season.

Collection and Habitat

Barberry often takes root in gaps, clearings, and young forests, sometimes forming dense stands throughout the understory that are impenetrable. It has been common in ornamental landscaping as hedgerows the past century and has escalated from there with the help of birds dispersing their seed.

It prefers partial sunlight but will survive shade and is considered invasive throughout the northeastern United States from North Carolina and Tennessee north to Nova Scotia and Canada and as far west as Montana.

Medicinal Uses

Parts used: root bark, fruit, leaf, and stem

TCM: Xiao bo (*Berberis chengii*); bitter, cold; associated with the stomach, large intestine, liver, gallbladder, and heart

Used to treat cancers or tumors of the liver, neck, or stomach and lumps of the liver. Fruits are used to relieve itching and other skin ailments, as well as to sweeten the breath and to promote general health.

Ayurveda: Daaruharidraa, Daruharidra, Daarvi (*Berberis aristata*), Himalayan barberry

Classically, barberry is used for anemia, jaundice, diseases of the mouth, swollen gums, ulcers, piles, skin diseases, conjunctivitis, leprosy, dysurea and other genitourinary disorders, dysentery, cholera, and malaria.

Western botanical: Used as a bitter tonic and for treatment of dyspepsia, liver disease, and diarrhea and as a mild purgative for constipation, fevers (typhus), and sore mouth and throats. Ancient Egyptians mixed the berries with fennel seed to treat fevers.

Plant Chemistry

Acetic acid, aluminum, ascorbic acid, berbamine, berberine, berberubine, berculcine, bervulcine, beta-carotene, caffeic acid, calcium capsanthin, chelidonic acid, chlorogenic acid, chromium, chrysanthemumxanthin, citric acid, cobalt, columbamine, columbianine, flavoxanthin, fructose, glucose, hydrastine, iron, isotetrandrine, jattrorrhizine, kaempferol, lutein, magnesium, magnoflorine, malic acid, manganese, niacin, oxyacanthine, palmatine, pectose, phosphorus, potassium, quercetin, rutin, riboflavin, selenium, silicon, sinapic acid, sodium, tannin, tartaric acid, thiamin, tin, vulvracine, yatroricine, zeaxanthin, zinc

Pharmacological Actions

Antimicrobial, anti-inflammatory, antioxidant, antirheumatic, astringent, hemostatic, anticonvulsant, immunostimulant, uterotonic, cardiovascular, antipyretic, lowers cholesterol, acts as a cholagogue (stimulates secretion of bile and bilirubin for liver cirrhosis), is used for arthritis and rheumatism (roots) and as a diuretic, expectorant, and laxative (berries). Promotes the flow and discharge of bile into the small intestine.

Scientific Studies

The component berberine has been studied extensively and has been found to be most effective for gastrointestinal infection.

Antimicrobial: Berberine has shown in vitro effectiveness at inhibiting *Staphylococci, Streptococci, Salmonella shigella, Entamoeba histolytica, E. coli, Vibro cholera, Giardia,* and *Candida albicans.* Although barberry has been used traditionally for remittent-type malarial fevers, berberine has been proved to not be a curative agent in malaria. Instead, it appears the compound liberates the parasites into the blood circulation, provoking the latent malaria and helping to diagnose it. Blood films screened prior to berberine administration were negative, with those afterward positive.

Antioxidant: Berbamine has shown itself not only to assist white blood cells when overloaded with toxic chemotherapy and radiation, but also to exhibit scavenging effects on free radicals and to perform growth-inhibiting actions on tumor cells with in vivo and in vitro studies. Berberine was also found to protect the liver from damage resulting from excessive acetaminophen intake.

Anticancer: Many studies show a couple of compounds present in barberry have strong anti-inflammatory and antitumor virtues. Berbamine was found to have very high antitumor activity against leukemia cells in vitro. Berberine was said to effectively inhibit COX-2 transcriptional activity in colon cancer cells in one study, while also inhibiting carcinogenesis and protecting the livers of small animals injected with the cancer-provoking chemicals 20-methylcholanthrene or N-nitrosodiethylamine.

Antihypertensive: Barberry water extract has shown antihypertensive and vasodilating effects on hypertensive rats.

Other Uses

The nutritious berries contain vitamins, minerals, sugars, organic acids, and pectin and have been used to make excellent jams, preserves, sour sauces, cold drinks, and liqueurs that at one time were common in Europe. The very young, tender leaves may also be eaten and added to salads. The roots (boiled in lye) and young leaves have been used as a dye for their orange-yellow coloring.

Ecological Importance

In forest clearings, barberry creates dense stands of entwining, thorny branches that protect overthinned forests, stabilize disturbed land, and allow the area to recover. Its berries provide abundant food for wildlife, and it provides nectar and pollen for insects. Barberry is a medicinal substitute for the endangered berberine-rich plants goldenseal and goldenthread, and it overlaps their regions of habitation.

Harvesting and Preparing

Harvest and prepare as described for using internally or externally.

Dosage

Medium-dose botanical.

Decoction: Ayurvedic and TCM use 1 to 3 grams or ¼ teaspoon of powdered bark

Tincture: 10 to 30 drops

Cautions and contraindications: Contraindicated during pregnancy. High doses may interfere with vitamin B metabolism.

Herb and drug interactions: Cyclophosphamide—an in vitro study showed berberine was effective in preventing cyclophosphamide-induced cystitis in rats. It helps block the effects of the drug, including bladder edema and hemorrhage, and other effects of cyclophosphamide urotoxicity.

References

Z. Fatehi-Hassanabad, M. Jafarzadeh, A. Tarhini, and M. Fatehi, "The Antihypertensive and Vasodilator Effects of Aqueous Extract from *Berberis vulgaris* Fruit on Hypertensive Rats," *Phytother Res* 19, no. 3 (2005): 222–25.

G. Y. Wang, Q. H. Lv, Q. Dong, et al., "Berbamine Induces Fas-mediated Apoptosis in Human Hepatocellular Carcinoma HepG2 Cells and Inhibits Its Tumor Growth in Nude Mice," *J Asian Nat Prod Res* 11, no. 3 (2009): 219–28.

Xie J, Ma T, Gu Y, Zhang X, Qiu X, Zhang L, Xu R, Yu Y. "Berbamine derivatives: A novel class of compounds for anti-leukemia activity," *European Journal of Medicinal Chemistry* 44, no. 8 (2009): 3293–98.

K. Fukuda, Y. Hibiya, M. Mutoh, M. Koshiji, S. Akao, H. Fujiwara, "Inhibition by Berberine of Cyclooxygenase-2 Transcriptional Activity in Human Colon Cancer Cells," *J Ethnopharmacol* 66, no. 2 (1999): 227–33.

A. H. Amin, T. V. Subbaiah, and K. M. Abbasi, "Berberine Sulfate: Antimicrobial Activity, Bioassay and Mode of Action," *Can J Microbiol* 15 (1969): 1067–76.

T. V. Subbaiah and A. H. Amin, "Effect of Berberine Sulphate on *Entamoeba histolytica*," *Nature* 215 (1967): 527–28.

D. Sun D, H. S. Courtney, and E. H. Beachey, "Berberine Sulfate Blocks Adherence of *Streptococcus pyogenes* to Epithelial Cells, Fibronectin, and Hexadecane," *Antimicrob Agents Chemother* 32(1988): 1370–74.

G. H. Rabbani, T. Butler, J. Knight, et al., "Randomized Controlled Trial of Berberine Sulfate Therapy for Diarrhea Due to Enterotoxigenic *Escherichia coli* and *Vibrio cholerae*," *J Infect Dis* 155 (1987): 979–84.

Y. Kumazawa, A. Itagaki, M. Fukumoto, et al., "Activation of Peritoneal Macrophages by Berberine-type Alkaloids in Terms of Induction of Cytostatic Activity," *Int J Immunopharmacol* 6 (1984): 587–92.

C. W. Wong, W. K. Seow, J. W. O'Callaghan, Y. H. Thong, "Comparative Effects of Tetrandrine and Berbamine on Subcutaneous Air Pouch Inflammation Induced by Interleukin-1, Tumour Necrosis Factor and Platelet-activating Factor," *Agents Actions* 36 (1992): 112–18.

H. S. Ju, X. J. Li, B. L. Zhao, et al., "Scavenging Effect of Berbamine on Active Oxygen Radicals in Phorbol Ester-stimulated Human Polymorphonuclear Leukocytes" *Biochem Pharmacol* 39 (1990): 1673–78.

Bindweed
Convolvulus arvensis

Common names: bindweed, field bindweed, wild morning glory

Family: Morning Glory family (Convolvulaceae)

Related species: *Convolvulus* subspecies is comprised of twenty-one species, mainly found around the Mediterranean, including *C. scammonia* and hedge bindweed (*Calystegia sepium*).

States where bindweed is considered invasive

The Plant

According to the traditional information gathered from around the world where these creeping plants thrive, *Convolvulus* species are inherently known as purgatives—that is, they cleanse or rid an organism of impurities. Some sinners live the afterlife in purgatory in order to cleanse themselves of guilt, sin, or ceremonial defilement: this showy plant, with a close allegiance to the divine psychotropic heavenly blue morning glory (*Ipomea purpurea*), brings peace to the heart and can transport us to heavenly realms and free us of impurity. By helping to untangle the mind's knots, bindweed can also tranquilize those who think excessively and tend to be manic. In the environment, bindweed inhabits, integrates, and aerates disturbed soil and improperly managed or otherwise stressed agricultural lands.

Its name comes from the Latin *convolvulus,* meaning "to twine around," and *arvensis,* meaning "cultivated fields." Bindweed's importance lies in its ability to clean a wide range of heavy-metal contaminants, which are often due to the overuse of chemical fertilizers and pesticides, thereby helping to purge chemically laden, overused farmland. For people, *Convolvulus* helps rid the body of toxins that accumulate in it; it specifically influences the liver, intestines, and central nervous system.

Description

Bindweed is a twining perinennial vine, with the notable funnel-shaped morning glory flowers (1–2 in. diameter) that vary in color from white to pale pink. The flowers open with glory by the rise of the sun in the morning, and then close in evening, when, as if sleeping, their petals shrink and twist back around themselves. Flowers give rise to tiny light brown fruits ($^1/_8$ in.) that contain two seeds. Bindweed's arrow-shaped leaves (1–2.5 in.) alternate around the vine that emerges from a root that can extend up to 18 feet in any direction.

Collection and Habitat

Originally dispersed throughout Eurasia, bindweed first appeared in North America in the 1730s, sold as an ornamental plant and medicine and also accidently spreading through agricultural seed shipments and ending up as primarily an agricultural weed. By the end of the 1800s, it was considered naturalized and was reported throughout the continental United States and Canada. It is now considered a pest throughout the world. It shows up in disturbed areas and appears along roadsides, field edges, streams, in shrubs, and on grasslands. Once it shows up, it is difficult to eradicate.

Medicinal Uses

Parts used: root, whole plant

TCM: No species of *Convovulus* is used in TCM.

Ayurveda: Shankhapushpi, Shankhaahvaa (*C. microphyllus, C. pluricaulis*), whole plant

Classically, bindweed is used as a brain tonic and stimulant to promote intellect and alleviate insanity, epilepsy, confusion, insomnia, psychoneurosis, and neurological disorders. It is also used as a tranquilizer and blood purifier, for uterine bleeding, spermatorrhoea, ulcers, bleeding piles, and venereal diseases.

Western botanical: Bindweed's medicinal actions were first described by the Greek physician Dioscorides, who used the plant to stop internal bleeding and to help heal wounds. Other early uses included

treating fevers and as a purgative. It is said to produce laxative qualities similar to jalap (*Ipomoea jalapa*), but with effects that are much
less drastic; when mixed with honey, it is even appropriate as a laxative for children. In *A Modern Herbal,* however, "Meyrick states that
the root of *C. arvensis* is a rough purgative, and to such constitutions
as can bear it, will prove serviceable in jaundice, dropsy and other
disorders arising from the obstructions of the viscera,"[4] and according to Rosemary Gladstar, "the roots are excellent for diarrhea."[5] The
Native Americans made use of bindweed as a cold infusion. It was
taken and used as a lotion for spider bites, to treat swallowing a spider, and to treat excessive menstruation. South Americans considered
the whole plant to be a laxative, and the leaves were used to stimulate
bile flow.

Plant Chemistry
Pseudotropine, tropine, tropinone, meso-cuscohygrine, cuscohygrine,
hygrine, calystegines, *n*-alkanes, *n*-alkanols, A-amyrin, campesterol, stigmasterol, B-sitosterol, aspartic acid, cystine, alanine, arginine

Pharmacological Actions
Purgative, anticancer medicine, antitumor medicine, anti-inflammatory,
antidiabetic medicine, cholagogic, and tranquilizing medicine

Scientific Studies
Anticancer: An extract of *C. arvensis* was found to inhibit tumor growth
in vitro by roughly 70 percent and improved lymphocyte survival ex
vivo, all with the use of nontoxic doses.

Antidiabetic: *C. arvensis* has showed significant inhibitory activity
toward beta-glucosidase and alpha-galactosidase, which helps decrease
the absorption of carbohydrates in the intestines for type 2 diabetes,
similar to the alpha-glucosidase inhibiting medications Precose (acar-
bose) and Glyset (miglitol).

Antimicrobial: In studies, *C. arvensis* was found to have broad-spectrum
antimicrobial and antifungal activity in vitro, with the flower being
most potent. The affected microorganisms were *Sarcina maxima,*
Staphylococcus aureus, Escherichia coli, Salmonella sp., *Bacillus subtilis, Micrococcus Kristina, Moraxella lucunata, Mycobacter phely,
Pseudomonas auregenosa,* and *Candida albicans.*

Antistress: Clinical trials of *C. pluricaulis* demonstrated its ability to reduce different types of stress of psychological, chemical, and traumatic origins, and it produced significant tranquilizing effects.

Other Uses

Though not often considered a food plant, the roots and young shoots of *C. sepium* are said to have been cooked as food in some parts of Asia, and the shoots and tender stems of sea bindweed were pickled in vinegar in the Old World. Bindweed is in the same family as *Ipomoea batatas,* the sweet potato, which is rich in vitamins and minerals and also, similar to sweet potato, contains an insulin-like compound that makes it useful in the management of diabetes. The plant is also used for making ropes, weaving, and netting.

Ecological Importance

In a phytoremediation study, bindweed has been found to clean chromium and copper and is considered a candidate to hyperaccumulate cadmium from contaminated soils. It was also found to be one of a few plant species to survive in an area full of heavy metals (Pb, Cu, Zn, Cd, Tl, Sb, and As) after the toxic spill of the Aznalcóllar mine in Spain and as well found assisting in the degradation of fuel oil from the Prestige oil spill in Spain. After finding bindweed growing in these areas, it was proposed as a remediator for the sites.

Through the compounds called calystegins found within the roots, bindweed acts as a unique kind of nutritional mediatior that provides an exclusive carbon and nitrogen source to the rhizosphere bacteria, thereby influencing and enriching the soil in a similar fashion to some nitrogen-fixing plants.

In addition, a number of bird species eat the seeds of bindweed.

Harvesting and Preparing

Harvest the aboveground plant in the summertime, including the leafy growth and flower, and the root in the autumn. Prepare fresh or dried.

Dosage

Medium- to low-dose botanical.

Decoction: The ayurvedic dosage is 3 to 6 grams of the powdered whole plant.

Cautions and contraindications: Use with caution with antipsychotic, tranquilizing medications.

Herb and drug interactions: None known.

References

J. L. Gardea-Torresdey, J. R. Peralta-Videa, M. Montes, G. de la Rosa, and B. Corral-Diaz, "Bioaccumulation of Cadmium, Chromium and Copper by *Convolvulus arvensis L.*: Impact on Plant Growth and Uptake of Nutritional Elements," *Bioresour Technol* 92, no. 3 (2004): 229–35.

A. S. Awaad, M. H. Nawal, and A. F. Khadiega, "Alkaloids, Some Constituents and Antimicrobial Activity of *Convolvulus arvensis*," *Egyptian J. Desert Res* 54, no. 2 (2004): 315–26.

B. O. Sowemimo and N. R. Farnsworth, "Phytochemical Investigation of *Convolvulus arvensis* (Convolvulaceae)," *Journal of Pharmaceutical Sciences* 62, no. 4 (1972): 678–79.

F. A. Vega, E. F. Covelo, M. J. Reigosa, and M. L. Andrade, "Degradation of Fuel Oil in Salt Marsh Soils Affected by the Prestige Oil Spill," *J Hazard Mater* 166, no. 2–3 (December 6 2008): 1020–29.

Daniel F. Austin, "Bindweed (*Convolvulus arvensis*, Convolvulaceae) in North America, from Medicine to Menace," *Journal of the Torrey Botanical Society* 127, no. 2 (2000): 172–77.

R. J. Molyneux, Y. T. Pan, A. Goldmann, D. A. Tepfer, and A. D. Elbein, "Calystegins, a Novel Class of Alkaloid Glycosidase Inhibitors," *Arch Biochem Biophys* 304, no. 1 (1993): 81–88.

M. Del Río, R. Font, C. Almela, D. Vélez, R. Montoro, and A. De Haro Bailón, "Heavy Metals and Arsenic Uptake by Wild Vegetation in the Guadiamar River Area after the Toxic Spill of the Aznalcóllar Mine," *J Biotechnol* 98, no. 1 (2002): 125–37.

X. L. Meng, N. H. Riordan, J. J. Casciari, et al., "Effects of a High Molecular Mass *Convolvulus arvensis* Extract on Tumor Growth and Angiogenesis," *P R Health Sci J* 21, no. 4 (2002): 323–28.

Blackberry
Rubus spp.

Common names: blackberry, Himalayan blackberry, Asian blackberry, bramble (*Rubus discolor* or *R. armeniacus*).

Family: Rose family (Rosaceae)

Related Species: sand blackberry (*R. coneifolius*), shrubby blackberry (*R. fructicosus*), cutleaf blackberry (*R. laciniatus*), purple flowering raspberry (*R. odoratus*), black raspberry (*R. occidentalis*), and dewberry (*R. trivialis*).

States where blackberry is considered invasive

The Plant

Blackberry is a friend of mine, though we still have our discrepancies. She has taught me much about myself, and I have grown to appreciate her constant reminders. When my wife and I moved onto this piece of land, it had been recently logged, and a couple of acres had been cleared for a house site. Blackberry colonized this desolate hillside and protected it from further soil erosion and intrusion. The plant descended on this land like a lion protecting her cub, showing her thorny fangs with ferociousness. It was angry, protective, reclaiming the wild, and there was no venturing onto this space without bloody repercussions. As our relationship deepened, it became clear to me that blackberry's presence is my medicine and, in a way, brought me here so that I could work with it. The plant's appearance in my life served as an impetus to encounter directly deep disruptions within myself. The plant's rampant existence outside my door reflected the latent, brewing angers that stirred inside of me—the strong feelings I had about the forces in the world that take over landscapes and mindscapes.

> *You know the kind: the omnipresent power-holding individuals and corporations who drop sweet fruit to entice us, yet are prickly at the core.*

The nature of blackberry conveys to me the blood-sucking perversions of society that continue to take over and harm the environment in illogical, ever-expansive ways. The plant became a symbol for me of that which has allowed all these devastating effects around the world to take place. My therapy came in the form of ripping up these representatives of my anger, and the plant was glad to be of service, for I could transform my latent feelings into actions and use that energy in a productive manner. In this way, blackberry has been my medicine and has helped me release these deep-seated tensions, and as my tensions relaxed, the plant opened the way for gardens to fill the hillside.

Knots of the mind eventually spread into the body and translate as bodily tensions. Physical exertion is an essential component of the exhausting process of consciously unwinding our overtightened bundles of emotion. The intertwining nature of blackberry teaches us how to observe when we get stuck and how to respond. Strong reactions do not help the situation, and our slow and patient unwinding of the knotted parts is essential to freedom. My annual interaction with blackberry provides reminders of the influences in and around my life that tighten me up and make me mad, thereby allowing me to be more present to these pressures and emotions and to deal with them more effectively, efficiently, and consciously.

*Although blackberry can introduce itself with a bristly attitude,
in the end, it does give sweet fruit.*

Description

The thorny canes with clumps of black berries make this plant easy to remember. The canes can grow up to 10 feet and longer and generally grow into dense thickets if left unchecked. The foreign varieties of blackberry usually have five leaflets compared to three on native varieties, although sometimes North American bramble can have five-lobed leaves as well. White flowers with five petals emerge, and soon follows the slow-ripening of fruit ($^1/_2$–1 in.) from green to red and finally to black, abundant in the mid to late summer.

Collection and Habitat

Blackberry grows all over the world, particularly in the subarctic and subalpine regions where there is at least ten inchs of rainfall. This plant was first publicly introduced to the United States in 1885, when Luther

Burbank bred the "Himalayan Giant," and became widespread in nurseries and in agricultural experiment stations. Now it can be found almost anywhere there is an opening, sometimes up to five hundred canes per square yard. With its spreading roots, it quickly takes over, and its abundant fruit appeals to hungry foragers.

Medicinal Uses

Parts used: roots, leaves, and fruit. The root bark is most astringent and said to be high in tannins.

TCM: Fu Pen Zi (*Fructus Rubi*), "overturned bowl fruit," fruit; sweet, sour, slightly warm; associated with kidneys and liver

1. Tonifies kidneys, consolidates essence, and restrains urine.
2. Tonifies the liver and improves vision.

Used to treat frequent urination, spermatorrhea, impotence, diabetes (low-glycemic index), vision problems, dizziness.

Ayurveda: Various *Rubus* species used; similar to Western botanical uses.

Western botanical: The ancient Greeks used this as a remedy for gout. With their high tannin content, the astringent tea of roots, leaves, and fruit has been traditionally used for diarrhea, dysentery, stomach pain, gonorrhea, fever, back pain, as a female tonic, as a blood tonic, prolapse, and sore throat and mouth. Leaf washes are used for sores, ulcers, boils, piles, and pimples. A once-popular folk treatment involves eating the young shoots in a salad to fasten loose teeth. Approved by the German Commission E studies for diarrhea and inflammation of the mouth and pharynx.

Plant Chemistry

Arbutin, ascorbic acid, B-amyrin, boron, chlorogenic acid, citric acid, ellagic acid, ferulic acid, fupenzic acid, hydroquinone, inositol, isocitric acid, kaempferol, lactic acid, malic acid, neo-chlorgenic acid, oxalic acid, quercetin, rubinic acid, rubitic acid, rubusoside, sitosterol, stigmasterol, succinic acid, tannin, ursolic acid

Pharmacological Actions

Antimicrobial, antidiarrheal, astringent, anti-inflammatory, hormonal

Scientific Studies

Antimicrobial: Fu pen zi (*R. chingii*) fruit in a decoction was found to inhibit *Staphylococcus aureus* and *Vibrio cholera*. Blackberry leaf (*R. ulmifolius*) extract and its isolated polyphenols with antioxidant activity were found in vitro to inhibit *Helicobacter pylori* bacteria that are linked to stomach inflammation and ulcers.

Antioxidant: The seeds and seed oil, the fruit of multiple cultivars, and the extract of blackberry leaf have all been studied for antioxidant activity. Through analysis of the numerous antioxidant compounds found in the plant and in in vitro studies, blackberry was noted for high levels of radical scavenging, antiaging abilities. In one study, the extract exhibited potent radical scavenging activity comparable to that of alpha-tocopherol (vitamin E) used in cosmetics that protect against wrinkles and UV radiation. In another in vitro study, the antioxidant compounds in the juice of blackberry were found to be potent scavengers of peroxynitrite, a free radical that causes endothelial dysfunction and vascular failure.

Anticancer: In an in vitro study among various blackberry species, extracts of the fruit from a Jamaican *Rubus* species demonstrated moderate COX inhibitory activity and exhibited the greatest potential to inhibit cancer cell growth, inhibiting colon, breast, lung, and gastric human tumor cells.

Hormonal: The fruit of *R. chingii* was found to exert estrogen-like activity in rats and rabbits.

Other Uses

Blackberry can be used for fresh fruit eating (and young shoots can be boiled, rinsed, and eaten), jams, preserves, wine, beer, and vinegar (drunk for fevers). It is also useful as hair dye.

Ecological Importance

Blackberry accumulates zinc and copper from the soil. The plant takes up the arsenic-based herbicide MSMA into its fruit. Blackberry protects those places that have been ravaged by improper logging, clearing, or other human actions. Its presence in these areas helps minimize soil erosion and creates a barrier to allow the land to rest and rejuvenate. After providing cover for emerging saplings, over time the trees begin to shade out the blackberries, and they slowly retreat to areas that provide enough sunlight.

Blackberry provides good shelter for birds, mammals, and insects, and the flowers are an abundant source of nectar and pollen for bees and other insects. The fruits also provide a wealth of food for wildlife; more than one hundred fifty animal species consume it.

Harvesting and Preparing

Harvest all parts as available throughout the growing year. Prepare as described previously, fresh or dried.

Dosage

Medium- to high-dose botanical.

Decoction: The TCM dosage for fruit (dried) is 3 to 10 grams. Use 10 to 20 grams of root.

Tincture: 30 drops

Cautions and contraindications: The overripe fruits are said to be indigestible (they also contain fewer antioxidants), and excess can cause diarrhea.

Herb and drug interactions: None known.

References

A. C. Anderson, A. A. Abdelghani, J. Hughes, et al., "Accumulation of MSMA in the Fruit of the Blackberry (*Rubus* sp.)," *Journal of Environmental Science and Health, Part B* 15 (1980): 247–58.

S. Martini, C. D'Addario, A. Colacevich, et al., "Antimicrobial Activity against *Helicobacter pylori* Strains and Antioxidant Properties of Blackberry Leaves (*Rubus ulmifolius*) and Isolated Compounds," *International Journal of Antimicrobial Agents* 34, no. 1 (April 20, 2009): 50–59.

M. Herrmann, S. Grether-Beck, I. Meyer, et al., "Blackberry Leaf Extract: A Multifunctional Anti-aging Active," *International Journal of Cosmetic Science* 29, no. 5 (October 2007): 411.

U. Erturk, C. Yerlikaya, and N. Sivritepe, "In Vitro Phytoextraction Capacity of Blackberry for Copper and Zinc," *Asian Journal of Chemistry* 19, no. 3 (May/June 2007): 2161–68.

H. Jiao, and S. Y. Wang, "Correlation of Antioxidant Capacities to Oxygen Radical Scavenging Enzyme Activities in Blackberry," *Journal of Agricultural Food Chemistry* 48, no. 11 (November 2000): 5672–76.

I. Serraino, L. Dugo, P. Dugo, et al., "Protective Effects of Cyanidin-3-O-glucoside from Blackberry Extract Against Peroxynitrite-induced Endothelial Dysfunction and Vascular Failure," *Life Sciences* 73, no. 9 (July 18, 2003): 1097–114.

C. S. Bowen-Forbes, Y. Zhang, and M. G. Nair, "Anthocyanin Content, Antioxidant, Anti-inflammatory and Anticancer Properties of Blackberry and Raspberry Fruits," *Journal of Food Composition and Analysis* 22, no. 7–8 (2009): 627–766.

Dandelion
Taraxacum officinale

Common names: dandelion, common dandelion, lion's tooth

Family: Aster family (Asteraceae)

Related species: There are nine related native species to common dandelion that grow in various regions of North America, including California dandelion (*T. californicum*), rock dandelion (*T. laevigatum*), and harp dandelion (*T. lyratum*).

States where dandelion is considered invasive

As my child plays with dandelion and blows the furry head, I am transfixed by the plant's life. It is a miniature globe of infinite possibilities, and perhaps the seeds will float to faraway lands and nestle into the perfect spot . . . which happens to be just about anywhere it wants.

The Plant

Local farmer Peter Dunning of Springfield, Vermont, who provides fresh milk for my family, shared a great story of dandelion and how he used it for his cows. A couple of years back, in early spring, the cows were just getting out to pasture. Green roughage was just growing, and Peter expected richer and creamier milk from the cows and was disappointed with the amount of cream his cows were making. (The cream is usually about two inches thick on a half-gallon jar.) He began to notice there was no dandelion in the area where the cows pastured, so he decided to feed them dandelion by hand. By the next day, with this additional food, the cows' milk production expanded, and it thickened with cream—even beyond Peter's expectations. After he shared this story with me, I was reminded of the fact that this herb has been used by humans for millennia specifically to increase milk production in nursing mothers.

Have you ever noticed the milky sap that flows in the stems of dandelions?

I recall the common sight of lawnscapers bending over their manicured grasses with their spray bottles of herbicide to stop dandelions from growing. I know, though, that the plant will keep smiling with its sunshine face, soak up the chemicals as food, and pop up its head in a new place in the lawn.

Dandelion grows just about anywhere there are people. It is especially good at breaking up the green-grass monoculture, aerating the soil, and adding essential food for pollinators in a flowerless lawn. Too often, however, dandelion is punished for existing. Nevertheless, it is quite a patient plant, accepting all our insults yet growing nonetheless.

In phytoremediation programs, dandelion is known to clean the soil of the heavy metals copper, zinc, manganese, lead, and cadmium, and is capable of accumulating deep benefitial nutrients for the other plant life. In traditional and contemporary herbal medicine, dandelion is used to aid the detoxification processes of the liver and kidneys and has been shown to help rid the body of pathogenic influences. In addition to providing a rich source of potassium, dandelion contains a vast array of vitamins, minerals, and phytochemicals, and it helps balance trace elements within the body. Dandelion passes the criteria for chelation therapy of heavy metals by containing various antioxidants and assisting liver and kidney function. In addition, dandelion provides trace minerals that can bind and replace metals that cause toxicity, and it nourishes the blood with essential nutrients.

Dens leonis—the lion's tooth—cuts sharp into the land and clamps down upon toxic elements. It sends its taproot deep (like teeth chewing food) to gather nutrients, then brings them to the surface in order to revive the soil and provide for surrounding species. Likewise, the plant provides deeply penetrating medicine for invasive pathogens and toxic elements. In its own way, the plant is somewhat like the bacteria it fights: widespread and resistant to human assault.

Description

A plant that most everyone knows, whether as a city dweller or country folk, dandelion can easily be found by its bright yellow flowers and its downy, globe-shaped seed head. In the late spring through fall, the aster-like flowers (1.5 in.) emerge from tooth-shaped basal leaves (4–8 in.) upon a single hollow stem usually 6 to 8 inches tall.

Collection and Habitat

Originally from Eurasia, the ubiquitous dandelion is now found throughout northern temperate regions, inhabiting all of North America. It can be found green and flowering throughout much of the year, florishing in lawns, gardens, roadways, sidewalks, and almost anywhere else.

Medicinal Uses

Parts used: root, leaf, and flower

TCM: Pu gong ying (*Herba Taraxacum mongolicum*), herb; bitter, sweet, cold; associated with the liver, stomach

1. Clears heat, eliminates toxins
2. Promotes liver detoxification and benefits the eyes
3. Clears damp heat jaundice
4. Reduces abscesses and dissipates nodules
5. Promotes urination

Classically, dandelion is used to treat breast abscesses, sores, deep-set boils, redness and swelling of the eyes, jaundice, and painful urination.

Ayurveda: Kanaful, Kaasani Dashti, Hindbaa-al-Barri (*T. officinale*), root

Classically, dandelion is used to treat chronic disorders of the liver and skin, visceral diseases, and as a mild cholagogue, diaphoretic, and diuretic, and in chronic skin diseases.

Western botanical: Dandelion has been used extensively as medicine and is considered a general tonic and stimulant to the liver, kidneys, and urinary system. Dandelion has been used for liver disease, jaundice, kidney and gallstones, gravel, and constipation. The bitter component helps stimulate appetite and aid digestion.

Plant Chemistry

Amylase, androsterol, aneurine, apigenin-7-glucoside, arabinose, arnidiol, arsenic, ascorbic acid, asparaginic acid, barium, B-carotene, B-sitosterol, boron, bromine, caffeic acid, calcium, caoutchouc, cerotic acid, chlorine, choline, chromium, chrysanthemumxanthin, cluytianol, cobalt, copper, coumestrol, cryptoxanthin, cycloartenol, D-glucuronic acid, faradiol, flavoxanthin, fructose, germacranolide, glucose, glutamic acid, glycerol, homoandrosterol, homotaraxasterol, inulin, iodine, iron, lactucerol, lead, lecithin, levulin, levulose, linoleic acid, lutein, luteolin, magnesium, manganese, mannitol, melissic acid, molybdenum, mucilage, niacin, nickel, nicotinic acid, oleic acid, p-coumaric acid, palmitic acid, phlobaphene, phosphorus, pollinastanol, potassium, pseudotaraxasterol, riboflavin, rubidium, saccharose, saponin, selenium, silicon, sodium, stigmasterol, strontium sucrose, sulfur, tannin, taraxacerine, taraxacine, taraxacoside, taraxanthin, taraxasterol, taraxerol, taraxin acid, taraxol, tartaric acid, thiamin, tin, titanium, tyrosinase, violaxanthin, xanthophylls, zinc

Pharmacological Actions

Antibacterial, antiviral, antiparasitic, immune enhancing, hepatoprotective, diuretic, cholagogic

Scientific Studies

Antimicrobial: Dandelion has shown antimicrobial effects in vitro against *Staphylococcus aureaus,* B-hemolytic streptococcus, *Diplococcus pneumoniae, Diplococcus meningitides, Corynebacterium diphtheriae, Pseudomonas aeruginosa, Bacillus dysenteriae,* and *Salmonella typhi.*

Anticancer: One in vitro study evaluating the anticarcinogenic activity of *Taraxacum officinale* showed that a water extract of dandelion leaf decreased the growth of breast cancer cells and blocked the invasion of prostate cancer cells. In contrast, extracts of dandelion flower and root appeared to have no effect on the growth of either cell line, yet the root could block the invasion of breast cancer cells.

Antioxidant: The flavonoids and coumaric acid derivatives from dandelion flower possessed marked antioxidant activity in both biological and chemical models. The compounds showed DPPH-radical-scavenging activity, and they significantly inhibited peroxyl-radical-induced intracellular oxidation of cells.

Anti-inflammatory: In one study, 120 to 180 grams of pu gong ying dandelion herb in decoction relieved acute tonsillitis in 82 of 88 patients. Another report showed dandelion herb (with isatidis root ban lan gen) 98 percent successful in treating 45 patients with acute sore throat. An herbal paste of dandelion also showed effectiveness in treating 51 patients with infected burns.

Anti-obesity: Dandelion has been found to have inhibitory activities against pancreatic lipase in vitro and in vivo and was determined to be of use as a natural weight-loss agent.

Other Uses

The young leaves of dandelion are a rich source of vitamins, minerals, and nutrients and are excellent greens to eat raw, steamed, or added to soups. For millennia, people have consumed them, and they are especially favored by the French, who carefully cultivate superior salad varieties. The older leaves are tougher and more bitter than the younger leaves, and they are not recommended for fresh consumption, but they can be dried for winter tea. One may keep a healthy patch of this wild and medicinal food for fresh eating throughout the year by plucking flower buds as they emerge and cutting back the young nutritious leaves for regular salad mix. In colder climates, in late autumn, cover the plants with an ordinary flowerpot and put down a deep mulch to protect the plant through the winter. The early emerging greens will be a welcome addition to the diet and will help celebrate the arrival of spring. With spring consumption (as when they are most cherished by foragers in search of fresh greens), dandelion also begins cleansing the digestive system that endured a winter diet of rich and preserved food.

Dandelion can also be used to create wine or beer. Dandelion coffee is a nutritious substitute for regular coffee, and it provides essential vitamins and minerals and helps rehabilitate the kidneys and liver that can be burdened with excessive, long-term coffee consumption. To prepare: clean the roots, cut into pieces, quickly dry, then roast the roots in an oven. Grind as coffee and prepare as you would coffee. Add the ground roots in place of coffee entirely, or dilute coffee with the dandelion preparation, or combine the preparation with other herbs, such as roasted chicory root, carob powder, chocolate beans, and chaga mushroom.

Dandelion is extremely valuable to the honey industry. It blooms in the early spring and helps bees after their fruit tree–pollen harvest no matter how cold it may get. Dandelion then keeps blooming until late autumn, feeding the colonies of bees after the summer's abundance of flowers and helping them to store honey for the winter.[6]

Ecological Importance

In a study throughout Montreal's urban sites, dandelion was used to evaluate the bioavailability of trace metals. Dandelion was used for its ability to take up the following metals in soil: copper, zinc, manganese, lead, and cadmium. The plant was found growing throughout various sites, including abandoned industrial areas with high concentrations of metals in the soil.

Dandelion has also been used as a soil enhancer. It is considered an excellent dynamic accumulator of potassium, phosphorous, calcium, copper, and iron, bringing these subsurface minerals to the topsoil and thereby providing fertilizer that benefits the surrounding plants.[7]

It has been reported that nearly one hundred different kinds of insects visit dandelion for its nectar, and it provides important early spring and late autumn honey and nutrients for overwintering bees and insects.

Harvesting and Preparing

Harvest fresh dandelion leaf in the springtime to impart stronger actions to cool, detoxify, and assist the liver. The root is ideally harvested in the late autumn, when the energy is returning underground for the winter hibernation and the root is filled with starchy reserves. Prepare, as usual, fresh or dried, for teas and tinctures internally and as washes or salves topically.

Dosage

High-dose food-grade herb

Decoction: The TCM dosage is 9 to 30 grams of herb. Use same for root.

Tincture: 30 to 60 drops

Cautions and contraindications: In a few people, the milky sap can cause contact dermatitis. Internally, it is a very safe food-grade herb, but it can cause mild diarrhea in excess (as just about anything can in excess).

Herb and drug interactions: It has been proposed that the use of dandelion and quinolone antibiotics (ciprofloxacin) together may influence the bioavailability of antibiotics.

References

K. Marr, H. Fyles, and W. Hendershot, "Trace Metals in Montreal Urban Soils and the Leaves of *Taraxacum officinale*," *Canadian Journal of Soil Science* 79, no. 2 (May 1999): 385–87.

S. C. Sigstedt, C. J. Hooten, M. C. Callewaert, et al., "Evaluation of Aqueous Extracts of *Taraxacum officinale* on Growth and Invasion of Breast and Prostate Cancer Cells," *Int J Oncol* 32, no. 5 (2008): 1085–90.

J. Zhang, M. J. Kang, M. J. Kim, et al., "Pancreatic Lipase Inhibitory Activity of *Taraxacum officinale* In Vitro and In Vivo," *Nutr Res Pract* 2, no. 4 (2008): 200–203.

H. J. Jeon, H. J. Kang, H. J. Jung, et al., "Anti-inflammatory Activity of *Taraxacum officinale*," *J Ethnopharmacol* 115, no. 1 (2008): 82–88.

C. Hu, and D. D. Kitts, "Dandelion (*Taraxacum officinale*) Flower Extract Suppresses Both Reactive Oxygen Species and Nitric Oxide and Prevents Lipid Oxidation In Vitro," *Phytomedicine* 12, no. 8 (2005): 588–97.

English Ivy
Hedera helix

Common names: English ivy, gum ivy, true ivy, woodbind

Family: Ginseng family (Araliaceae)

Related species: Irish ivy (*H. hibernica*), quite similar in appearance to English ivy but botanically different, Persian ivy (*H. colchica*). Not related to poison ivy (*Toxicodendron radicans*) or Boston ivy (*Parthenocissus tricuspidata*)

States where English ivy is considered invasive

The Plant

Ivy is a terrifying force to the psyche, with its climbing tentacles that envelop buildings and trees. Ivy persistently looks for cracks to reach into. It has been known to tear buildings down, reach into the slats, pry open joints, and dig into masonry. Ivy has also been found to clean contaminants from the polluted air in its surroundings. This is a powerful plant in many ways, and similarly for human physiology it reaches into the farthest areas of the body, reduces the inflammation of toxic buildup, binds the deposits of toxins from the tissues, and helps eliminate these influences. The plant has been esteemed since ancient times, forming wreaths and crowns for poets and the noble, as well as a long history of use as a pain-relieving and detoxifying medicine. During this time it was bound around the brow to prevent intoxication. It is especially noted for topical use on neuralgia and arthritis and for skin afflictions such as sunburn, cancer, warts, and tumors. Internally, it is used to treat lung complaints. Ivy possesses numerous vitamins, antioxidants, and unique compounds, all of which support the health of an individual and address various disease conditions. Some of the known antioxidants within this plant include caffeic acid, campesterol, chlorogenic acid, quercetin, rutin, kaempferol, and alpha-hederin, all of which contribute to the anticancer, anti-inflammatory, and antimicrobial abilities of this plant, as well as the further use of ivy for invasive disease and toxic accumulation.

> *Though I've not used English ivy myself, my teacher Juliette de Bairacli Levy used it often for treating animals. She often told the tale of a flock of sheep that were mysteriously dying. She gave them all fresh ivy and the flock lived. I know this story to be true, because even years later after this event happened, the farmer whose sheep she cured with English ivy would write her words of thanks. Read in her book,* The Complete Handbook for Farm and Stable, *for some of her ivy stories.*

ROSEMARY GLADSTAR

Description

Ivy is a powerful climbing vine that suctions itself with its tendrils to buildings, trees, and anything else in its path and is well known for enveloping the "Ivy League." The evergreen leaves are dark and waxy and generally three-lobed with a heart-shaped base 1.5 to 2 inches across. As the plant matures and grows farther upward, touching sunlight, it begins to branch out and tends to have unlobed, rhomboid-shaped leaves. After reaching the sun the plant will then flower, producing clusters of small, greenish white flowers in the summer. In the late summer, small, purplish black, fleshy fruits appear with a seed or two within.

Collection and Habitat

English ivy grows throughout the temperate world, at first inhabiting Europe, western Asia, northern Africa, and now on the loose in the eastern and western margins of North America. Ivy was first introduced into the United States in 1727 when colonists were encouraged to grow it for mainly decorative reasons: to provide a green cover for both the ground and for buildings. It is still found throughout the city landscape on buildings and trellises, and now further distributed throughout disturbed forests. It prefers shady or semishady and moist areas, but it can tolerate drought.

Medicinal Uses

Parts used: leaves and berries

TCM: Chang chun teng (*Hedera nepalensis*); cold; associated with lungs, liver. Classically, *H. nepalensis* is used to treat skin diseases with inflammation.

Ayurveda: No known uses found in ayurveda.

Western botanical: Used to treat cancer, arthritis, sclerosis, swellings, sunburn, dysentery, jaundice, toothache, tumors, callus, corns, and warts.

Plant Chemistry

A-hederin, A-tocopherol, arabinose, bayogenin, B-elemene, B-sitosterol, E-hederin, caffeic acid, campesterol, chlorogenic acid, D-galactose, elixen, emetine, falcarinone, falcarinol, fat, germacrene, hederacosides, hederagenin, hederine, isochlorogenic acid, isoquercitrin, kaempferol, linoleic acid, malic acid, methylethylketone, methylisobutylketone, oleanolic acid, oleic acid, p-coumaric acid, palmitic acid, petroselenic acid, polyacetylenes, protocatechuic acid, quercetin, resin, rhamnose, rosmarinic acid, rutin, scopolin

Pharmacological Actions

Antibacterial, antiviral, antimycotic, anti-inflammatory, antiseptic, analgesic, astringent, cathartic, diaphoretic, emetic, emmenagogue, laxative, vasoconstrictor, vasodilator, vermifuge, anthelmintic, mollusicidal, antiflagellate, antispasmodic, expectorant

Scientific Studies

Antioxidant: In various studies, compounds found in the leaf and berries of English ivy and *H. colchica* exhibited strong antioxidant and metal chelating activity, comparable to, and sometimes exceeding, the abilities of vitamin E and other antioxidants.

Anti-inflammatory: Extracts of English ivy were found to possess anti-inflammatory effects in rat models, with effects seen in both acute and chronic cases of inflammation. In separate clinical studies, including one double-blinded, placebo-controlled trial, ivy leaf extract preparations showed effectiveness at improving the respiratory function of children with chronic asthma. In a postmarketing study of 9,657 patients

(5,181 children) with bronchitis (acute or chronic bronchial inflammatory disease), after seven days treatment with a syrup containing dried ivy leaf extract 95 percent of the patients showed improvement or healing of their symptoms with little incidence of adverse reactions (2.1 percent, with mainly GI disturbance).

Anticancer: In ongoing anticancer testing, the saponin extracted from English ivy, alpha-hederin, demonstrated in vitro antimutagenic activity against a clastogenic agent, doxorubicin, and an aneugenic agent, carbendazim. The saponins were also found to be cytotoxic to cultured B16-melanoma cells.

Antimicrobial: The saponin alpha-hederin was also found in vitro to inhibit the growth of protozoa and *Candida albicans* and was able to kill liver flukes both in vitro and in vivo.

Other Uses

Ivy has been used for tanning leather.

Ecological Importance

In the city environments where trees are far out-numbered by cars, pollution and smog fills the air both on the street and inside buildings. English ivy has stepped in, by persuading people to plant it, and provides powerful air-cleansing abilities for the urban setting. In phytoremediation studies, English ivy was found to be quite effective in removing indoor and outdoor volatile organic air pollutants including benzene, toluene, and PCBs.

The vine and its berries provide some bird species shelter and food.

Harvesting and Preparing

Harvest and prepare ivy throughout the year as described. For external application, prepare as a salve.

Dosage

Low- to medium-dose botanical, internally. Externally, ivy is considered safe.

Cautions and contraindications: Fresh leaves can be an irritant; there may be allergic symptoms from skin, eyes, and respiratory tract upon touching and inhaling.

Herb and drug interactions: None known.

References

M. H. Yoo, Y. J. Kwon, K. C. Son, et al., "Efficacy of Indoor Plants for the Removal of Single and Mixed Volatile Organic Pollutants and Physiological Effects of the Volatiles on the Plants," *Journal of the American Society for Horticultural Science* 131, no. 4 (2006): 452–58.

C. Moeckel, G. O. Thomas, J. L. Barber, et al., "Uptake and Storage of PCBs by Plant Cuticles," in *Environ Sci Technol* 42, no. 1 (January 2008): 100–105.

I. Gülçin, V. Mshvildadze, A. Gepdiremen, and R. Elias, "Antioxidant Activity of Saponins Isolated from Ivy: Alpha-hederin, Hederasaponin-C, hederacolchiside-E and Hederacolchiside-F," *Planta Med* 70, no. 6 (2004): 561–63.

H. Süleyman, V. Mshvildadze, A. Gepdiremen, et al., "Acute and Chronic Antiinflammatory Profile of the Ivy Plant, *Hedera helix,* in Rats," *Phytomedicine* 10, no. 5 (2003): 370–74.

I. Gülçin, V. Mshvildadze, A. Gepdiremen, and R. Elias, "The Antioxidant Activity of a Triterpenoid Glycoside Isolated from the Berries of *Hedera colchica:* 3-O-(beta-D-glucopyranosyl)-hederagenin," *Phytother Res* 20, no. 2 (2006): 130–34.

D. Hofmann, M. Hecker, and A. Volp, "Efficacy of Dry Extract of Ivy Leaves in Children with Bronchial Asthma: A Review of Randomized Controlled Trials," *Phytomedicine* 10 (2003): 213–20.

P. Villani, T. Orsière, I. Sari-Minodier, G. Bouvenot, and A. Botta, "*In vitro* Study of the Antimutagenic Activity of Alphahederin" *Ann Biol Clin* (Paris) 59, no. 3 (2001): 285–89.

S. Fazio, J. Pouso, D. Dolinsky, et al., "Tolerance, Safety and Efficacy of *Hedera helix* Extract in Inflammatory Bronchial Diseases under Clinical Practice Conditions: A Prospective, Open, Multicentre Postmarketing Study in 9657 Patients," *Phytomedicine* 16, no. 1 (2009): 17–24.

J. Moulin-Traffort, A. Favel, R. Elias, et al., "Study of the Action of Alpha-Hederin on the Ultrastructure of *Candida albicans,*" *Mycoses* 41, no. 9–10 (1998): 411–16.

S. Danloy, et al., "Effects of Alpha-hederin, a Saponin Extracted from *Hedera helix,* on Cells Cultured *In Vitro,*" *Planta Med* 60, no. 1 (1994): 45–49.

C. Moeckel, G. O. Thomas, J. L. Barber, et al., "Uptake and Storage of PCBs by Plant Cuticles," *Environ Sci Technol* 42, no. 1 (2008): 100–105.

Garlic Mustard
Alliaria petiolata

Common names: garlic mustard, hedge garlic

Family: Mustard family (Brassicacea)

Related species: There are 107 genera in Brassicaceae, and garlic mustard is the one species in *Alliaria*.

States where garlic mustard is considered invasive

The Plant

A nutritious food and herb, garlic mustard provides a wild source of powerful compounds. According to James Duke, *Alliaria petiolata* "may combine the activities and phytochemicals of the garlic and mustard families, e.g., respectively allicin and sulforaphane," and "the late Herb Pierson was looking at . . . Garlic-mustard . . . for [his] new Designer Food cancer-preventive program."[8] The plant is said to contain more vitamin A than spinach and more vitamin C than orange juice and to impart the medicinal virtues of both garlic and mustard that mainly focus on the digestive and respiratory systems.

> *I've used this herb in wild food dishes, added to soups, and/or used as a tea. It's delicious, pungent, and garlic flavored—so good to use wherever one would use mustard greens and/or garlic. It is warming and energizing, effective for deep bronchial inflammation and colds.*
>
> ROSEMARY GLADSTAR

As for environmental concerns, contrary to the common verbiage that trumpets garlic mustard's harm to surrounding life-forms, a study conducted by the department of biology at Boston University found that garlic mustard could actually provide a benefit to northeastern U.S. forests. According to the report, the soil in which garlic mustard flourished was "consistently and significantly higher in N, P, Ca and Mg availability, and soil pH,"[9] the soil nutrients that present conditions for optimal plant growth. Garlic mustard was found not to release volatile compounds from the roots to affect other plants and "did not alter soil nutrient cycling."[10] In addition, the leaves of garlic mustard significantly accelerated the decomposition of leaf litter from native trees to further the nutrient cycle and nourish the soil. With this, the report concluded that garlic mustard "creates a positive feedback between site occupancy and continued proliferation,"[11] and by feeding the soil with its presence, the plant helps nourish the other plant life within the vicinity of its effects.

Description

Garlic mustard is named for the pungent odor of its heart-shaped leaves. After germinating in the spring, the plant forms into a rosette the first year and then bolts with stalks and flowers (1–3 ft.) the second spring. Leaves alternate, and clusters of small white flowers ($^1/_4$ in.) form, which then give way to thin seedpods (1–2 in.) that split open and reveal small black seeds. The plants then die back after the seeds are dispersed in late summer and fall. The young plant resembles several native plants including violets (*Cardamine* spp.) and ground ivy (*Glechoma hederacea*).

Collection and Habitat

Garlic mustard is a European native and is now found throughout temperate North America after being introduced by settlers in the early 1800s. The plant inhabits most states except in the far south and southwest desert, growing in forests and along the forest edges, in disturbed areas, roadsides, trails, and along riverbanks.

Medicinal Uses

Parts used: whole plant
TCM: No known use of *Allaria* in TCM.
Ayurveda: No known use of *Allaria* in ayurvedic medicine.
Western botanical: According to Maud Grieve in *A Modern Herbal,*
  garlic mustard leaves are taken internally to promote perspiration and

clear phlegm from the lungs, and are applied externally on gangrene and ulcers for its antiseptic properties.[12] The seeds have also been used as snuff to excite sneezing.

In 1955, *The Dispensatory of the United States* (USD) reported that "the herb and seeds have been used for reputed diuretic, diaphoretic, and expectorant effects, and were also used as an external stimulant and postulant application."[13] This makes it an appropriate remedy for bronchitis, asthma, skin infections, itching, and sores. It has also been used for arthritis, neuralgia, and gout.

Plant Chemistry

Alliaroside, allyl-glucosinolate, 2-butyl-glucosinolate, allyl-isothiocyanate, allyl-sulfide, linoleic acid, linolenic acid, arachidic acid, ascorbic acid, cardeonlides, carotene, diallyl-sulfide, eicosadienic acid, eicosenoic acid, eo, erucic acid, feruloyl-choline, glycerol, glycolic acid, isothiocyanates, linoleic acid, linolenic acid, myrosin, oleic acid, palmitic acid, saponarin, sinapic acid, sinapine, sinigrin, squalene, stearic acid, tetracosadienoic acid

Pharmacological Actions

Antiasthmatic, antioxidant, antimicrobial, antiscorbutic, antiseptic, deobstruent, diaphoretic, sternutatory, vermifuge, vulnerary

Scientific Studies

In one study, garlic mustard seed extract was assessed for free radical scavenging components. Of forty-five different plant species studied, *Alliaria petiolata* was one of the top three plants to provide potent antioxidants.

Other Uses

Garlic mustard is a delicious, nutritious green with high values of vitamins A and C that complements any fine dish. It can be added to soups, the leaves can be stuffed, it can be made into pesto, and it can be used as any pot herb or wild green we eat as a vegetable. Garlic mustard can be eaten fresh in salads or as a cooked green, and the dried herb can be taken as a tea or as a nutritious flavoring for winter meals.

A yellow dye can be made from the whole plant.

Ecological Importance

As stated in the plant introduction, it can greatly benefit northeastern forests in phytoremediation. Garlic mustard also serves as a general nectar-producer for bees, butterflies, moths, and flies in the forest understory.

Harvesting and Preparing

Harvest in the spring and early summer, dry and store the extra for adding to soups as a green enhancement in the wintertime.

Dosage

High-dose herbal food.

Cautions and contraindications: None.
Herb and drug interactions: None known.

References

V. L. Rodgers, B. E. Wolfe, L. K. Werden, et al., "The Invasive Species *Alliaria petiolata* (Garlic Mustard) Increases Soil Nutrient Availability in Northern Hardwood-conifer Forests," *Oecologia* 157, no. 3 (September 15, 2008): 459–71.

Y. Kumarasamy, M. Byres, P. J. Cox, et al., "Screening Seeds of Some Scottish Plants for Free Radical Scavenging Activity," *Phytotherapy Research* 21, no. 7 (July 2007): 615–21.

James Duke, "Phytomedicinal Forest Harvest in the United States," www.fao.org/docrep/W7261E/W7261e17.htm (accessed March 18, 2009).

Japanese Honeysuckle
Lonicera japonica

Common names: honeysuckle, lonicera

Family: Honeysuckle family (Caprifoliaceae)

Related species: coral honeysuckle (*Lonicera, L. confusa, L. fuchsioide, L. involucrate, L. lanceolata*)

States where Japanese honeysuckle is considered invasive

Oh sweet nectar, soothing inflammations, and cooling our tempers . . .

The medicine enters the mouth, relieving the parched tongue of its thirst, and then with swallowing, it coats the sore throat with cooling virtues.

The honeysuckle flower sticks out her tongue at you, and, as you look down her throat, the red stem reveals her medicine.

The Plant

Honeysuckle (*Lonicera* spp.) has been used since time immemorial for the detoxifying nature of its medicine. Modern science now verifies these abilities by isolating numerous compounds that have been identified with antioxidant, antimicrobial, and anti-inflammatory natures and outlines a perfect remedy for our present-day ills and inflictions.

Japanese honeysuckle has recently received bad press for harboring Lyme disease–bearing ticks, but upon closer examination, we see that this plant might very well be powerful treatment against the disease. Traditionally, in Asian medicine, *Lonicera japonica* flowers and stems have been used to treat fevers; infection; skin rashes; and arthritic, painful, swollen joints—

and these are the classic symptoms of Lyme disease. Japanese honeysuckle has been found to possess powerful, broad-spectrum action against a variety of microbes. In addition, in one study, *Lonicera japonica* was the primary herb in a formula that successfully treated syphilis in most sufferers. The syphilis pathogen is a close relative of the Lyme spirochete; finding plants that can relieve syphilitic conditions can help lead to treatment of Lyme infection. With this strong relevance to one invading disease and long use as a medicine against various infections, *Lonicera* will likely also prove benefitial to other invading and antibiotic-resistant pathogens that are sneaking into our surroundings.

Description

Japanese honeysuckle is a climbing vine with sweet-scented, white to pink flowers that fade to yellow and bloom from late spring into fall. The leaves are often lobed at the base of the vines and then become simple, oval, and opposite farther up the stem (1.5–3.5 in.). Japanese honeysuckle can be distinguished from native honeysuckle species by the black fruits that mature in the fall, whereas natives have red to orange fruit.

Collection and Habitat

Originally from eastern Asia, Japanese honeysuckle was brought to the United States in 1806 and soon spread far and wide through the nursery trade as well as for promotion of wildlife habitat and erosion control. Japanese honeysuckle is now found throughout the eastern states, growing in disturbed areas, roadsides, forest openings, and field edges. Its range is kept in check by severe winter temperatures and low precipitation.

Medicinal Uses

Parts used: flower, leaf, stem, and fruit

TCM: Jin yin hua, "golden silver flower," flower; Ren dong teng, "tolerate winter vine," stem; sweet, cold; associated with the lungs, stomach, and large intestine

1. Clears heat.
2. Eliminates toxins.
3. Clears damp heat diarrhea.

The stem is similar to the flower, yet it is milder and has an affinity for the joints and limbs of the body. Classically, Japanese honeysuckle is used to treat fevers, sore throat, laryngitis, enteritis, heat stroke, dysentery, diarrhea (with blood and/or mucus), sores, boils, abscesses, lesions, scabies, eczema, measles, carbuncles, ulcerations, itching, painful swellings, tumors, venereal disease. In seventh-century China, honeysuckle was advocated for treatment of diabetes mellitus, which, it was then noted, was often accompanied by skin infections.

Ayurveda: No known uses.

Western botanical: There are eclectic uses of the syrup made from the flowers of *L. caprifolium* for asthma and other respiratory disorders, and the juice is used for the relief of bee stings. Native Americans used *L. involucrate* as a lactagogue, laxative, poultice, and vermifuge, and for sore, swollen, and aching legs and feet.

Plant Chemistry

Lonicerin, loniceraflavone, loganin, luteolin, luteolin-7-rhamnoglucoside, inositol, tannin, chlorogenic acid, isochlorogenic acid, aromadendrene, linalool, geraniol, octanal, A-pinene, B-pinene, myrcene, B-terpinene, 1,8-cineole, hexenol, A-terpinene, neral, linalyl acetate, geranial, citronellol, terpinyl acetate, eugenol, B-eudesmol, geranyl acetate, A-copene, patchoulene, A-caryophyllene, B-caryophyllene, iso-bornyl acetate, farnesol, nerolidol, caffeic acid, rutin, isoquercitrin, 1-O-caffeoylquinic acid, kaempferol, oleanolic acid, quercetin

Pharmacological Actions

Antibacterial, antiviral, antifungal, antitumor, antioxidant, antiinflammatory, antipyretic, central nervous system stimulant, antihyperlipidemic, gastrointestinal, antidiarrhetic, diuretic, hypo- and hyperglycemic, emeto-cathartic, antispasmodic

Scientific Studies

Antimicrobial: Japanese honeysuckle flower has been found to have strong, broad-spectrum inhibitory effects in vitro against *E. coli, salmonella typhi, Pseudomonas aeruginosa, Staphylococcus aureus, B-hemolytic streptococcus, Streptococcus pneumonia, Mycobacterium tuberculosis, Bacillus dysenteeria, Vibrio cholera, Diplococcus pneumonia,* and *Diplococcus meningitides.* A study of a Chinese herbal formula containing Japanese honeysuckle and other herbs (including smilax, licorice, and dandelion) found it successful in the treatment of syphilis. Using blood tests, 90 percent of those with acute syphilis proved negative, as well as 50 percent of those with chronic cases.

Antioxidant: Numerous antioxidant compounds have been found in Japanese honeysuckle. They have been identified as chlorogenic acid, caffeic acid, quercetin, rutin, isoquercitrin, luteolin, lonicerin, kaempferol, 1-O-caffeoylquinic acid, dicaffeoylquinic acid, oleanolic acid, and protocatechuic acid. At least one company is isolating chlorogenic acid from honeysuckle flower and making a concentrated medicine with it. Currently, there are many studies that delineate numerous health attributes of the above constituents.

Other Uses

The flowers make a pleasant tea that is consumed widely in Asia and is especially helpful in cooling individuals in the summer heat. The flowers also contain a few drops of delicious, sweet nectar that draws many children. They are also ideal for children's heat rash and toxic sores. The Japanese parboil the young leaves and eat them as a vegetable. Some species (generally, the native ones) provide fruit that is edible fresh or cooked and that contains pectin and sugars. Hybrids produce large, flavorful fruits, though the Japanese honeysuckle is not edible and should be considered toxic for eating purposes.

Ecological Importance

Japanese honeysuckle is another viney weed that webs itself throughout disturbed landscapes, often entangling the edges of clearings and not allowing further entrance. Honeysuckle flowers provide sweet nectar for hummingbirds and act as a general source of nectar for insects and bees. The plant also serves as shelter for various creatures.

Harvesting and Preparing

Harvest the flower just prior to its opening in the early morning and lay it out to dry in the sun. Once spread out, the flowers should not be handled, for they are delicate and might cause a dark stain. The stems are harvested in the fall and winter and dried in bundles with leaves removed. According to some, soaking the fresh leaves in water produces a stronger antibiotic action than decoction, though I would include the flowers as well. The herb can be charred to enhance the plant's ability to stop bleeding in cases of bloody diarrhea.

Dosage

Medium- to high-dose botanical.

Decoction: The TCM dosage is 10 to 20 grams of the flower, with smaller doses used for colds and flu, and up to 60 grams taken for severe heat toxin. Use 10 to 30 grams of the stem in decoction.

Cautions and contraindications: None known.

Herb and drug interactions: None known.

References

C. W. Choi, H. A. Jung, S. S. Kang, et al., "Antioxidant Constituents and a New Triterpenoid Glycoside from Flos Lonicerae," *Arch Pharm Res* 30, no. 1 (2007): 1–7.

D. Tang, H. J. Li, J. Chen, et al., "Rapid and Simple Method for Screening of Natural Antioxidants from Chinese Herb Flos Lonicerae Japonicae by DPPH-HPLC-DAD-TOF/MS," *Journ Sep Sci* 31, no. 20 (2008): 3519–26.

States where Japanese knotweed is considered invasive

Japanese Knotweed
Polygonum cuspidatum

Common names: Japanese knotweed, bushy or Chinese knotweed, Japanese bamboo, Mexican bamboo. Other Latin names used for this species include *Reynoutria japonica* and *Fallopia japonica*.

Family: Polygonaceae. In this family there are about 49 genera and 1,100 species worldwide. There are approximately 24 genera and 446 species in the United States and Canada, many of which have been introduced.

Related species: Pennsylvania knotweed (*Polygonum pensylvanicum* L.), giant knotweed (*P. sachalinense, P. hydropiperoides* Michx., *P. sagittatum* L., and *P. arifolium* L.)

The Plant

I just finished cutting up the Japanese knotweed root that I harvested yesterday, and now it's setting in the racks to dry. I'm exhausted with all of this work I've put into this harvesting, but at the same time, a subtle force has taken hold of me. Those who have dared to dig up the roots know how difficult it is to get to this plant. The roots on an established clump are massive, intertwined, and extremely strong. They take many whacks with my spade and sometimes all my arm strength before I can free them from their layered mass of collective plant root.

This is a warrior plant that has no fear of shovel, poison, or extreme environment. Any bit of root that isn't removed when digging—or that breaks off near a stream—will plant itself and establish large colonies if conditions are right. Knotweed is one of the cleansing plants, so it gravitates to polluted areas, and it continues to thrive when a chemical wash of Roundup comes its way. In its native Japan, it is a pioneer plant of the volcanoes, and for fifty years or so after it has moved in it is the dominant species until other plants can join it. Outside its homeland, Japanese knotweed is nearly exclusively a female plant, and the Japanese Knotweed Alliance has postulated, "In total biomass terms, this clone is probably the biggest female in the world!"[14] This means that most often, knotweed doesn't spread by seed in Europe and North America. Its roots' strength, ability to adapt, and tenacity in moving and establishing itself nearly anywhere is also reflected in the potency of its medicine for both the land and the people.

Knotweed is now widely used for the treatment of Lyme disease, and according to Stephen Buhner, the plant tends to move into infected areas six months to one year prior to the arrival of the disease. In addition to addressing a variety of invading pathogens with potent antimicrobial tendencies, Japanese knotweed's abundance of vitamins, minerals, and antioxidants help the body to process numerous toxins and cancers. One component, resveratrol, is produced in knotweed at greater concentrations than any other plant in the world and garners much attention (and hundreds of millions of dollars) with its antiaging virtues.

Just think—it grows everywhere for us.

Japanese knotweed in bricks

Description

This plant stands 6 to 10 feet tall, with bamboolike stalks. After overwintering, shoots appear from underground rhizomes early in the spring, soon after a good thaw. They grow rapidly and can exceed 2 inches of growth per day, which means that they attain full height by the early part of summer. Although leaf size varies (and leaves do not look like those of bamboo) they are normally about 6 inches long and 3 to 4 inches wide and broadly oval or pointed at the tip. The minute, greenish white flowers occur in attractive, branched sprays in late summer or early fall, and if they are pollinated, they are followed soon after by small, winged fruits. The seeds are triangular, shiny, and very small—about $^1/_{10}$ inch long. The aboveground parts of the plant are killed by the first frost, but stems remain standing into the next growing season.

Collection and Habitat

Native to Japan, northern China, Taiwan, and Korea, it was introduced into Britain in the 1840s as an ornamental, for fodder, and for erosion control, and it was introduced into America in the late nineteenth century. It quickly naturalized and is now established throughout North America, Europe, and New Zealand, creating anywhere from a couple of clumps of plants to dense acres wherever it occurs. This plant tolerates extreme conditions and grows from Alaska to Louisiana, in the volcanic fumaroles of Japan, in heavily contaminated soils, and especially along roadside ditches and streambanks. It generally grows as an herbaceous plant in wetland habitats. Knotweed tolerates a variety of soil types—including silt, loam, and sand—in soils with a pH range from 4.5 to 7.4, and in heavily contaminated soil. It has spread extensively solely by vegetative reproduction (by roots and root fragments), and the University of Leicester in Britain discovered that knotweed is not only a single sex but is also a single clone that grows throughout the British Isles and some of North America. It is possible that additional clones of the plant have crossed pollinated with other wild *Polygonum* species here in North America, as well as with imported male plants, and therefore have potentially reproduced and created viable seeds and offspring that have spread throughout.[15]

Medicinal Uses

Parts used: The root/rhizome is used for medicine, and the shoots are used for food.

TCM: Hu zhang, "tiger cane," *itadori* (Japanese: "heals the sick"); bitter, cold; associated with the liver, gallbladder, and lungs

1. Invigorates the qi and blood, stops pain
2. Clears heat, resolves toxins
3. Transforms phlegm, stops cough

Classically, knotweed is used to treat arthritis, traumatic injury, jaundice, urinary disorders, constipation, menstrual disorders, kidney stones, burns, hemorrhoids, carbuncles, skin infections, snakebite, dental caries

Ayurveda: Japanese knotweed is not traditionally used in India, but some other species of *Polygonum* are used in a way that's similar to how they're used in TCM.

Western botanical: Japanese knotweed is relatively new to the Western herbal world, but it has received considerable research in a roundabout way through the study of its components, most notably resveratrol, trans-resveratrol, polydatin, and emodin. It is now also receiving a great deal of attention for use in the treatment of Lyme disease, with the most extensive outline for use as an extremely important and highly potent medicine in Stephen Buhner's book *Healing Lyme*. In fact, by some regards, Japanese knotweed is the most important herb of choice for treating Lyme disease, specifically addressing the central nervous system and the neurological complications of infection, as well as killing of the Lyme bacteria itself.

Plant Chemistry

Barium, bromine, calcium, catechin, chrysophanol, citreorosein, copper, dimethylhydroxychromone, emodin, emodin monomethyl ether, fallacinol, glucofragulin, glucoside, iodine, iron, isoquercitrin, manganese, methylcourmarin, molybdenum, napthoquinone, nickel, phosphorus, physcion, physide, piceid, plastoquinone, polydatin, polydatoside, polygonin, potassium, protocatechuic acid, quercitrin, questin, questinol, resveratrol, reynoutriin, rheic acid, rubidium, rutin, sulfur, tannin, titanium, trans-resveratrol, and zinc

Pharmacological Actions

Antibacterial, antiviral, antispirochetal, antifungal, immunostimulant, immunomodulant, anti-inflammatory, central nervous system (brain and spinal cord) relaxant, CNS protectant, anti-inflamatory, antioxidant, antimutagenic, anticarcinogenic, angiogenesis modulator, vasodialator, antiasthmatic, cardioprotective, antiathersclerotic, antihyperlipidemic, antineoplastic, hepatoprotective, inhibits platelet aggregation, antithrombotic, antipyretic, analgesic, antiulcer, hemostatic, and astringent

Scientific Studies

Japanese knotweed contains resveratrol, the antioxidant of the "French Paradox," and praised for its existence in red wine. Knotweed is worthy of at least four patents on its constituents used for treating neurodegenerative diseases, and the entire herb contains many other wonder drugs, including trans-resveratrol, emodin, and polydatin. Resveratrol is most commonly known, and a vast array of studies have been performed with it. There are more than a hundred patents on the compound for treating a variety of conditions, including cancer, inflammations, and neurodegenerative diseases.

Antioxidant: One study conducted by the National Institutes of Health (NIH) demonstrated that resveratrol improved health and survival in old, overweight male mice. These mice, whose high-calorie diet was supplemented by resveratrol, had enhanced health and longer lives when compared to the aged, overweight mice who did not receive it. Japanese knotweed root extracts in another study in vitro have clearly shown antioxidant activity by protecting DNA from hydroxyl radical-induced DNA damage.

Antimicrobial: In vitro studies using the whole herb have shown broad antibacterial effects that inhibit the growth of *Staphylococcus aureus, S. albus, Neisseria cattarrhalis, α* and *β Streptococci, E. coli, Proteus vulgaris, Pseudomonas aeruginosa, Salmonella typhi,* and *Shigella flexneri.* Other studies using a 10 percent decoction of whole root have found broad antiviral effects against influenza (type A), enteric cytopathic human orphan (ECHO) virus, and herpes simplex viruses.

Anticancer: Research is proving the wonder-drug capabilities of resveratrol: the compound seems to inhibit the proliferation of many cancer cells while not harming the liver. In vitro studies show the antioxidant can aid pancreatic cancer treatment and cripple the function of pancreatic cancer cells while sensitizing them to chemotherapy. According to Paul Okunieff, chief of radiation oncology at the University of Rochester Medical Center, "resveratrol seems to have a therapeutic gain by making tumor cells more sensitive to radiation and making normal tissue less sensitive."[16] Data also suggests its beneficial use in breast cancer for its chemopreventive effects and in ovarian cancer (it significantly inhibits migration and adhesion of cancer cells).

Neurological: In neurodegenerative diseases, resveratrol activates a gene (SIRT1), which promotes neuron survival and can help prevent and treat Alzheimer's disease. The compound has also been found to protect neurological cells against dopaminergic neurodegenerative disorders such as Parkinson's disease. Dopamine, which occurs naturally in the brain and acts as an oxidant, in turn causes degeneration of neurons. Resveratrol, serving as an antioxidant, can protect these neurons. In a similar fashion, resveratrol might be beneficial for brain regeneration in cases of ischemia stroke and in recovery after brain injury. The compound polydatin, has demonstrated a neuroprotective effect on cerebral injury as well.

Hormonal: Trans-resveratrol is a phytoestrogen, a naturally occurring plant compound that shares many attributes of the female hormone estrogen. In how it mimics estrogen, trans-resveratrol could play a role in protecting against bone loss induced by estrogen deficiency. Among other things, estrogen stimulates cancer cells to grow and multiply, and trans-resveratrol has antiestrogenic properties that, on certain types of cancer cells, block receptors to which estrogen usually binds. In a sense, the compound fools cells into thinking that they already have estrogen. This property keeps estrogen away from cells, thus preventing the hormone's effects.

Wound healing: Japanese knotweed has long been used for treating burns and wounds. Studies of knotweed decoction have shown actions promoting eschar formation (scabbing) along with the exhibition of antiseptic effects.

Other Uses

If we reflect on what the Japanese Knotweed Alliance said (this plant is "the largest female in the world in biomass terms"), then we should explore Japanese knotweed's use as biomass fuel and its use in the creation of electricity (instead of diverting food crops to this industry). There's a great deal of knotweed everywhere, with the potential of annual or twice-annual harvests along convenient roadsides for biofuel opportunities. Developing business and government jobs to harvest and, in turn, use this plentiful plant for a community's electrical needs would be another excellent way to enhance local economies and work with what is on hand.

Also, the young, foot-long shoots of knotweed can be eaten and prepared like asparagus or rhubarb.

Ecological Importance

The prolific presence of this plant in roadside ditches and on other polluted and toxic land is a clear indicator of its ability to transform heavily contaminated soils. It has been shown to tolerate and actually clean soils contaminated with zinc, lead, and copper and is found throughout old copper and zinc mine areas and along notoriously polluted streams. If we begin to explore this plant's chemistry and see how these chemicals alter human bodies, then we can overlap human systems with natural ecosystems. It contains the compounds resveratrol, trans-resveratrol, polydatin, and emodin, which have been shown to possess antimicrobial, anti-inflammatory, antioxidant, and regenerative properties, and is very good at ridding the body of deep infection and toxins, especially in hard-to-reach places. It is no surprise, then, that it grows in such infected and toxic places.

Japanese knotweed is considered an excellent general nectar producer, attracting plenty of native bees and insects to its flowers.

Harvesting and Preparing

Get ready to get dirty, sweaty, and maybe even mad. Knotweed makes you work hard for its medicine, and that imbues you with a certain power, and the knotweed's use of its own warrior strength in the process of harvesting these mighty roots potentiates the medicine. Dig by hand or use a tractor for larger operations. Clean well and cut into manageable pieces, then dry.

To prepare for external use on burns, skin infections, hemorrhoids, and snakebite, use the fresh herb, soak a dressing or cotton ball in decoction, or use the powered herb and apply throughout the day on the affected area.

Dosage

Medium- to high-dose botanical.

Decoction: prepare by cooking 1 ounce of the herb with 1 liter of water. Simmer for 20 minutes, strain, cool, and drink in 4 equal doses throughout the day.

Tincture: 30 to 60 drops (Although, Buhner advises against using tinctures in treating Lyme disease.)

Cautions and contraindications: Contraindicated in pregnancy. Otherwise, Japanese knotweed is a very safe herb. Occasional side effects include digestive complaints such as gas and bloating, a bitter taste in the mouth, dry mouth, nausea, abdominal distention, and diarrhea. These are all signals to reduce the dosage.

Herb and drug interactions: Use caution with insulin and other anti-diabetic drugs because there could be a synergistic effect between any one of these and knotweed, creating hypoglycemia. It should not be used with blood-thinning medications, and discontinue use of the herb ten days prior to any surgery.

Resources

Japanese Knotweed Alliance, "What is Japanese Knotweed?" www.cabi.org/japaneseknot-weedalliance/Default.aspx?site=139&page=52 (accessed January 19, 2008).

Joan Spainhour, *Medical Attributes of* Polygonum cuspidatum—*Japanese knotweed* (Wilkes Barre, Pa.: Wilkes University, 1997), http://klemow.wilkes.edu/Polygonum.html (accessed April 25, 2008).

Z. P. Liu, W. X. Li, B. Yu, et al., "Effects of Trans-Resveratrol from *Polygonum cuspidatum* on Bone Loss Using the Ovariectomized Rat Model," *Journal of Medicinal Food* 8, no. 1 (March 1, 2005): 14–19; doi:10.1089/jmf.2005.8.14.

Y. Cheng, H. T. Zhang, L. Sun, et al., "Involvement of Cell Adhesion Molecules in Polydatin Protection of Brain Tissues from Ischemia-reperfusion Injury," *Brain Research* 1110, no. 1 (September 19, 2006): 193–200.

J. Baur, K. Pearson, et al., "Resveratrol Improves Health and Survival of Mice on a High-calorie Diet," *Nature* 444 (November 2006): 337–42.

P. Okunieff, W. Sun, W. Wang, J. Kim, and S. Yang, "Anti-cancer Effect of Resveratrol Is Associated with Induction of Apoptosis via a Mitochondrial Pathway Alignment," *Advances in Experimental Medicine and Biology* 614 (2008): 179–86.

L. Feng, L. F. Zhang, T. Yan, et al., "Studies on Active Substance of Anticancer Effect in Polygonum cuspidatum," *Zhong Yao Cai* 29, no. 7 (2006): 689–91.

D. Bagchi, D. K. Das, A. Tosaki, M. Bagchi, and S. C. Kothari, "Benefits of Resveratrol in Women's Health," *Drugs Exp Clin Res* 27, no. 5–6 (2001): 233–48.

C. Y. Hsu, Y. P. Chan, and J. Chang, "Antioxidant Activity of Extract from *Polygonum cuspidatum*," *Biol Res* 40, no. 1 (2007): 13–21.

Y. Kimura, and H. Okuda, "Resveratrol Isolated from *Polygonum cuspidatum* Root Prevents Tumor Growth and Metastasis to Lung and Tumor-induced Neovascularization in Lewis Lung Carcinoma-bearing Mice," *J Nutr* 131, no. 6 (2001): 1844–49.

C. Wang, D. Zhang, H. Ma, et al., "Neuroprotective Effects of Emodin-8-O-beta-D-glucoside In Vivo and In Vitro," *Eur J Pharmacol* 577, no. 1–3 (2007): 58–63.

P. Marambaud, H. Zhao, and P. Davies, "Resveratrol Promotes Clearance of Alzheimer's Disease Amyloid-beta Peptides," *J Biol Chem* 280, no. 45 (2005): 37377–82.

D. Delmas, B. Jannin, and N. Latruffe, "Resveratrol: Preventing Properties Against Vascular Alterations and Aging," *Mol Nutr Food Res* 49, no. 5 (2005): 377–95.

X. Z. Ding, et al., "Resveratrol Inhibits Proliferation and Induces Apoptosis in Human Pancreatic Cancer Cells," *Pancreas* 25 (2002): 71–76.

P. L. Krishna, D. L. Bhat, Christov Konstantin, et al., "Estrogenic and Antiestrogenic Properties of Resveratrol in Mammary Tumor Models," *Cancer Research* 61 (2001): 7456–63.

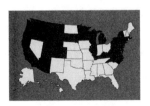

States where spotted
knapweed is considered
invasive

Knapweed or Star Thistle
Centaurea spp.

Common names: spotted knapweed (*C. maculosa* or alternative botanical names *C. biebersteinii* and *C. stoebe*), red star thistle, caltrops (*C. calcitrapa*), yellow star thistle, Barnaby's thistle (*C. solstitialis*), cornflower, bachelor's buttons, bluebonnet (*C. cyanus*)

Family: Aster family (Asteraceae)

Related species: Russian knapweed (*C. repens* or *Acroptilon repens*), diffuse knapweed (*C. diffusa*), Iberian star thistle (*C. iberica*), *C. nigra, C. jacea, C. nigrescens, C. trichocephala*

The Plant

Red star thistle (*Centaurea calcitrapa*) is named for the resemblance of its flower head to the ancient weapon caltrops, a spiked iron ball thrown to the ground to lame horses' feet during battle. The name still seems relevant today, because star thistle is at the front lines of the War on Plants in the rangelands of the Wild West, with the plant defending the land it has claimed against the ruthless livestock that have burdened the fragile soil.

Knapweed (*Centaurea* spp.) creates a dense stand in a knappy kind of way—entangling a web of spidery spindles and acting as one of the opportunistic fellows who place barriers to access an overburdened land. These plants are an expression of such an ecosystem, telling any who will listen about the imbalance and disturbance and sending deep roots through the hardpan soil to access essential elements that the land's original inhabitants can't reach, and then bringing them to the surface. Cornflower (*C. cyanus*) acts as a powerful deterrent and indicator to the farmer who has depleted the soil, and Grieve once wrote of "its tough stems in former days of hand-reaping were wont to blunt the reaper's sickle, earning it the name of Hurt-Sickle."[17]

The *Centaurea* species contain many novel and powerful constitutes, some so potent as to be considered poison to the surrounding plants and, in some cases, to livestock as well. Though common lore has it that these allelochemicals are produced to kill and mame the surrounding plant life, this is yet another human and scientific projection and is not based on a deeper ecological understanding. Indeed, these poisonous actions on the central nervous system of horses are being studied as potential treatment for Parkinson's disease, where in small doses these "poisons" act as anti-oxidants for both the land and for people.

> *The difference between medicine and poison is all about dose.*
> HIPPOCRATES

Description

Yellow star thistle (*C. solstitialis*) is a hardy annual, or sometimes biennial, that grows from 6 inches to 5 feet tall depending on the conditions. The plant is most notable for its bright yellow, thistlelike flowers with sharp, long spines protecting them. After emerging in the winter months, the dusky, silver-green shoots and leaves give way to flowers in early spring that continue to bloom until early the following winter. Basal leaves are deeply lobed and grow 2 to 3 inches, while the leaves along the flowering stalk are smaller (1 inch) and narrower, with no lobes. Yellow star thistle also sustains itself with a deep taproot that can grow 3 feet into the soil to collect moisture. Like yellow star thistle, red star thistle (*C. calcitrapa*) has flower heads that are defended by sharp spines, though its flowers are reddish purple in color. The plant has a thistlelike appearance as well, with sharp green leaves that are deeply lobed or sometimes divided. The leaves alternate around the multistemmed plant, and the young leaves and stems are covered with many small sticky hairs. It is difficult to distinguish from Iberian star thistle (*C. iberica*).

Spotted knapweed (*C. maculosa*) is easily identified by its purple bachelor's-button-like flowers blooming in the summertime. It grows as a biennial or short-lived perennial, first appearing as a rosette of deeply lobed leaves (8 inches long) and then producing flowering stalks (growing up to 3 feet) with smaller, silvery green leaves with fewer lobes that alternate around the stalk. After blooming, the seed head turns brown, with black tipped bracts on fringed edges, which serve as an identifying feature of this knapweed species.

Collection and Habitat

Yellow star thistle is of Middle Eastern, Mediterranean, North African, and Eurasian origins and likely immigrated to North America along with imported alfalfa seed stock. Other knapweeds also traveled across the oceans with seed stock and in livestock feed in the mid to late 1800s. Red star thistle comes from the region between the Black and Caspian seas, and spotted knapweed hails from Europe and western Asia. Now, people are the major cause of the spread of knapweed species throughout North America: the seeds sneak easily into hay, gravel, crop seed, and trucks. They spread and form dense stands throughout pasture and rangelands, floodplains, open forests, and along roadsides. Knapweed species are found throughout the United States and southern Canada. The star thistles most abundant in the western United States have long taproots to reach deep sources of water.

Medicinal Uses

Parts used: root, whole herb, flower, seed

TCM: No known use.

Ayurveda: Behman Safed (*C. behen*), white rhapontic root.

Classically, *Centaurea behen* is used to treat weakness of brain, heart, and liver, palpitation, hepatitis, melancholia, sexual debility, neurasthenia, spermatorrhoea, fatigue, and for diseases of the stomach and intestines. *C. picris* is used as a powdered herb in treatment of worms, and for livestock wounds.

Middle Eastern: Qanturyun aynata/shawk al-dardar (*C. ainetensis*), Qantaryun (*C. erengoides*)

Several *Centaurea* species are used in Turkish folk medicine to alleviate pain and inflammation in rheumatoid arthritis, high fever, and headache.

Western botanical: Eclectic practitioners considered the virtues of star thistle (*C. calcitrapa*) to be similar to those of blessed thistle (*Cnicus benedictus*), which was valued for its treatment in intermittent diseases, inflammation, fever, hepatitis, jaundice, arthritis, epilepsy, and externally for shingles outbreak. Other sources compare it to gentian in its use as a bitter tonic to improve digestion and to treat jaundice, fever, skin outbreaks, and infection.

According to Maud Grieve, "the seeds used to be made into powder and drunk in wine as a remedy for stone, and the powdered root was considered a cure for fistula and gravel."[18] From Timothy Coffey's book, Parkinson wrote in 1629 that the remedy of *Centaurea* is to use "not only against the plague and pestilential diseases, but against the poison of scorpions and spiders."[19] Culpeper wrote that knapweed "gently heals up running sores, both cancerous and fistulous, and will do the same for scabs of the head."[20]

Cornflower (*C. cyanus*) has been used for fever, constipation, leucorrhea, menstrual disorders, candida, as a laxative, tonic, diuretic, expectorant, or as a stimulant for liver and gallbladder function. It is also reported as a nervine that relieves discomforts caused by nerve impairment. Externally, the flower is used in eye washes for inflammation and conjunctivitis, and for eczema, sores, wounds, hemorrhoids.

More recently, yellow star thistle (*C. solstitialis*) is earning attention for causing equine nigropallidal encephalomalacia, a poisoning of horses (from eating an excessive quantity of the plant) that results in a Parkinson's-type disease. Originally, it was thought that this was caused by two potent neuroexcitotoxic amino acids, aspartic and glutamic acids, but other studies suggest that different mechanisms must be present to account for the varying levels of these components. Other amines, notably tyramine, have been found. This is of great interest for use in medicine, leading to studies that seek the connection to Parkinson's disease treatment.

Plant Chemistry

Cornflower, bachelor's button (*C. cyanus*)—acetic acid, aluminum, apigenin, apiin, cichoriin, cnicin, cyanidin-3,5-diglucoside, cyanin, cyanocentaurein, fragarin, iron, mucilage, pelargonin, polygalacturonic acids, protocyanin, quercimeretrin, tannin, trideca-tetrain, wax

Red star thistle (*C. calcitrapa*)—apigenin, arachidic acid, astragalin, behinic acid, b- amyrin, B-sitosterol, centaurin, choline, cichoriin, cniciin, inulin, lignoceric acid, linoleic acid, myristic acid, naringenin, oleic acid, palmitic acid, pectin, quercetin, rutin, scabiolide, stearic acid, stizolophine

Yellow star thistle (*C. solstitialis*)—aspartic acid, centaurocyanin, chlorojanerin, cyanidine, glutamic acid, solstitialin, succinylcyanin, tyramine

Pharmacological Actions
Antimicrobial, anti-inflammatory, anticancer, antioxidant, antiulcer, antinociceptive, antipyretic, and hypoglycemic

Scientific Studies
Antimicrobial: Various *Centaurea* species have shown in vitro broad antimicrobial activities, including inhibiting effects on some multiple antibiotic resistant bacteria, *Enterococcus faecalis,* and a DNA virus.

Anti-inflammatory: *C. ainetensis,* a plant endemic to Lebanon and used by Middle Eastern herbalists, contains bioactive compounds that have shown in vitro and in vivo anti-inflammatory and analgesic actions.

Anticancer: *C. ainetensis* has also shown that its extract and its bioactive molecule, salograviolide-A, can inhibit the proliferation of human colon cancer cells without harming the epithelial host cells. The sesquiterpene lactones from the aboveground parts of different *Centaurea* species have also displayed in vitro cytotoxic/cytostatic activity against at least three human cell lines.

Antiulcer: The fresh, spiny flowers of yellow star thistle (*C. solstitialis*) have long been used for peptic ulcers in Turkey, and in vitro studies have shown the components solstitialin-A and chlorojanerin help prevent lesions and inhibit the output of stomach acids in order to heal the ulcers.

Antioxidant: A study in Italy of one of their species found *Centaurea centaurium* to possess "high free radical–scavenging activity" and "potent antioxidant properties" due to various fatty acids and terpenes the plant produces.[21]

Other Uses
According to *The Encyclopedia of Edible Plants,* the young shoots of *C. calcitrapa* and other species have been eaten as food in Europe, Asia, and Africa.

Ecological Importance
Instead of finding that the allelopathic chemical in spotted knapweed (*C. maculosa*), catechin, causes oxidative stress to the surrounding plant life and thereby kills it, a USDA study found that this plant acted as a potent antioxidant to the soil, restored the balance of microbial life, and cleansed it of any free radicals or toxicities. A separate study also revealed

that other phytotoxins of *C. diffusa* facilitate nutrient uptake for the plant, helping it acquire iron from deep within the earth and providing the element to the surface soil. It is also true that iron is scarce in many of the deprived, alkaline-rich environments that knapweed likes to invade. Also, with deeply penetrating taproots 3 to 6 feet long, dense stands of the yellow star thistle (*C. solstitialis*) may restore the fertility, structure, and permeability of the degraded soils in which it inhabits.

It is not merely a plant; it is an expression of the land.

Further, the color of cornflower is said to change depending on the composition of the soil. Bachelor's button/cornflower (*C. cyanus*) is a hyperaccumulator of nickel and is a good candidate for the remediation of radiocesium (Cs-134). In addition, bachelor's button was found surviving—along with a handful of other species—in a site heavily contaminated with oil products and therefore considered for future phytoremediation by natural attunation. *C. virgata* was also determined to be a moderately high accumulator of iron, copper, zinc, and cadmium.

Because it is a hyperaccumulator, knapweed should be explored in phytomining the West of the heavy-metal-riddled land. Through the refining process, its biomass could also be used for energy.

Harvesting and Preparing

Harvest and prepare as described. Externally, it can be prepared as a wash for irritated eyes, sores, wounds, and shingles.

Dosage

Low- to medium-dose botanical.

Cautions and contraindications: Knapweed has been found to poison horses foraging on it.

Herb and drug interactions: None known.

References

R. D. Reeves and N. Adiguzel, "The Nickel Hyperaccumulating Plants of the Serpentines of Turkey and Adjacent Areas: A Review with New Data," *Turkish Journal of Biology* 32, no. 8 (2008): 143–53.

S. R. Tang and N. J. Willey, "Uptake of Cs-134 by Four Species from the Asteraceae and Two Varieties from the Chenopodiaceae Grown in Two Types of Chinese Soil," *Plant and Soil* 250, no. 1 (2003): 75–81.

S. Moreta, T. Populina, L. S. Contea, et al., "HPLC Determination of Free Nitrogenous

Compounds of *Centaurea solstitialis* (Asteraceae), the Cause of Equine Nigropallidal Encephalomalacia," *Toxicon* 46, no. 6, November (2005): 651–57.

R. S. Talhouk, W. El-Jouni, R. Baalbaki, et al., "Anti-inflammatory Bio-activities in Water Extract of *Centaurea ainetensis*," *Journal of Medicinal Plants Research* 2 (2008): 24–33.

F. Conforti, F. Menichini, M. R. Loizzo, et al., "Antioxidant, Alpha-amylase Inhibitory and Brine-shrimp Toxicity Studies on *Centaurea centaurium* L. Methanolic Root Extract," *Natural Product Research* 22, no. 16 (2008): 1457–66.

Kudzu
Pueraria montana var. *lobata*

Common names: kudzu vine

Family: Pea family (Leguminosae)

Related species: tropical kudzu
(*P. phaseoloides*)

States where kudzu is
considered invasive

The Plant

Kudzu (*Pueraria Montana*) is another invasive plant that has found the southern culture to be of its liking. Nicknamed "the vine that ate the South," this tenacious Asian native has trampled throughout the region, creating a tangled mess. Kudzu was originally introduced to America in 1876, and in the 1930s and '40s farmers were paid eight dollars per acre by the Soil Erosion Service to plant kudzu, and more than 1.2 million acres were planted throughout the southern United States under this subsidized program. The plant also proved to be agreeable fodder for the livestock, so, combined, modern-day human activities have led directly to the rampant spread of this plant. Expansion of infrastructure, abandonment of farmland, and the establishment of plantations of trees for the lumber industry are major contributors to helping move this plant throughout the area. Kudzu is a warrior plant that tries to protect these areas with its viney presence, creating a fencelike barrier to keep out further intrusion. This ability is exemplified by not even allowing battle tanks on a Virginia military base to move through their viney entanglement. Kudzu follows along the trail of petroleum that is left behind by massive fossil fuel–based machinery, and it helps clean this contaminant and other toxins from the environment as it protects it. This steadfast plant essentially counters the agro-military-industrial fossil fuel–based encroachments, and in order to do so, has to be a very powerful entity indeed.

> *Kudzu may, in fact, become the plant to inspire and sustain the southern culture's love of race cars and help keep NASCAR's engines running into the post-oil era by supplying the fuel of the future through its ethanol-producing roots. And kudzu can help with the hangovers after the race . . .*

As a medicine, *Pueraria montana* has been used for thousands of years to treat drunkenness and hangover, as well as for the onset of colds and flu accompanied by fever, headaches, skin rash, muscle stiffness, and pathogenic influences of the bowels. Kudzu is also a rich food source long used in Japanese cooking—it is full of valuable vitamins, minerals, and unique phytochemicals that have proved their antioxidant and detoxifying effects in modern study.

Description

Kudzu is a trampling vine that completely encases anything in its way: trees, buildings, powerlines, or hillsides. Each root can penetrate 10 feet (3 meters) into the ground and weigh a hundred pounds, while one root can produce thirty vines with each vine reaching 100 feet (30 meters). The leaves arrange themselves alternating around the stem, each leaf divided into three leaflets (4 inches wide), which may be lobed or unlobed. In midsummer, kudzu bloom long hanging clusters of fragrant, purple flowers about ½ inch long. These turn into brown, hairy, flat seedpods with 2 to 10 hard seeds.

Collection and Habitat

Kudzu is native to East Asia, Malaysia, Oceania, and the Indian subcontinent, and it has been successfully introduced to Australia, Switzerland, and North and South America, though only in the southern United States is kudzu considered a serious pest. It is now present throughout the eastern half of the United States and the Pacific northwest, and it is expected to continue to migrate northward with the current climate changes. Kudzu has tolerance to drought, frost, grazing, heavy soil, slope, vines, and other weeds, but it cannot survive a deep freeze, the tropics, sandy soils, or waterlogged fields.

Medicinal Uses

Parts used: Whole plant, root, leaf, and flower
TCM: Ge gen (*Radix Puerariae*); sweet, acrid, cool; associated with the spleen and stomach

1. Dispels wind, releases muscles, and clears heat
2. Nourishes fluids and alleviates thirst
3. Vents measles
4. Stops diarrhea

Classically, kudzu was used for more than two thousand years to treat headaches, fevers, and stiffness and pain in neck and shoulders. It promotes the eruption of measles and treats dysentery, burning or chronic diarrhea, hypertension, and alcoholism, and its flower is used for drunkenness or hangover.

Ayurveda: Vidaari Kanda (*Pueraria tuberosa*). Used similarly as TCM.
Western botanical: No known use.

Plant Chemistry

Arachidic acid, ash, B-sitosterol, calcium, carotene, daidzein, daidzin, eicosanoic acid, formononetin, genistein, hexadecanoic acid, irisolidon, iron, magnesium, p-coumaric acid, phosphorus, potassium, puerarin, quercetin, riboflavin, robinin, silica, tectoridin, tetracosanoid acid

Pharmacological Actions

Cardiovascular, antiplatelet, antihypertensive, antipyretic, antidiabetic, antispasmodic, anti-alcoholic

Scientific Studies

Anticancer: The isoflavones puerarin, daidzein, and genistein were analyzed for their ability to inhibit the growth of breast cancer cells in vitro. These compounds were found to exhibit anticancer abilities and inhibited the proliferation of breast cancer cells by reducing cell viability and inducing apoptosis. With this, it has been suggested for use as a chemopreventive and/or chemotherapeutic agent against breast cancer.

Antioxidant: Puerarin, the main isoflavone glycoside found in the root of kudzu, has been found to be a potent antioxidant that protects the liver from carbon tetrachloride (CC14) toxicity in mice studies. The antioxidant properties of puerarin were also studied for treatment in Alzheimer's disease. This isoflavone was found to protect against free radical damage to the brain by plaque that commonly forms in Alzheimer's disease. Puerarin was isolated and demonstrated to inhibit oxidative injury via estrogen receptor mechanisms.

Antiviral: The water extract of kudzu was effective at inhibiting the cytotoxic effect of enterovirus 71 infection that can cause brain encephalitis and death.

Heart protective: In studies for the treatment of prophylaxis and ischemic heart disease, an herbal preparation of puerarin (isolated from kudzu) and danshensu (isolated from *Salvia miltiorrhiza*) at a 1:1 ratio, was found to "exert significant cardioprotective effects against acute ischemic myocardial injury in rats,"[22] with anti-lipid peroxidation and antioxidant effects.

Alcoholism: In numerous studies, kudzu extract and in particular its isoflavones, puerarin, daidzin, and daidzein, have been shown to lessen cravings and reduce alcohol intake. In a placebo-controlled double-blind study, fourteen heavy beer drinkers significantly reduced their volume consumed and the size of each sip they swallowed after they took 1 gram of kudzu extract 3 times daily for a week. Another empirical trial showed that 80 percent of alcohol abusers no longer experienced cravings.

Antihypertensive: Treatment of 222 patients with hypertension accompanied by neck stiffness and pain resulted in 78–90 percent rate of effectiveness using kudzu root.

Migraine headaches: Anecdotal evidence suggests that kudzu may prove useful in managing cluster headache.

Other Uses

Kudzu has been introduced throughout the world for a variety of economic and practical benefits. The leaves, shoots, and roots of kudzu can be eaten, and a starch is obtained from the tuberous root. The roots are a rich source of carbohydrates, fat, fiber, protein, vitamins, and minerals. In China and Japan, Ko-fen flour, made from the unprocessed root, is ground with water and then allowed to settle in order to extract the starch. The starch is then used as a thickener for soups and is combined with millet to make a nourishing and medicinal porridge. The kudzu powder is especially useful for excess heat, with fever, irritability, and dehydration, due to the root's cooling nature and ability to generate fluids. Kudzu is also said to be cultivated for its tuber in the uplands of New Guinea and New Caledonia.

Kudzu was introduced as fodder, and it was grown for pasture, hay, and silage in the southern states. It is palatable to all types of livestock and is nearly as nutritious as alfalfa.

It was employed originally as a fast-growing, energy-saving cooler and was planted around buildings to provide shade. A useful fiber for weaving is obtained from stems.

It is being studied as a potential source of biofuel. Kudzu root that weighs up to one hundred pounds each and that contains as much as 50 percent starch by weight is ideal for ethanol production. The vines can grow a foot a day, and they are an abundant source of biomass for energy. This creates financial incentives and is a productive way to rid the South of this overabundant plant.

Ecological Importance

In phytoremediation studies, Kudzu has been found to clean contaminated sites, by removing petroleum and chromium from the soil. In the early 1900s, kudzu was first employed for erosion control and soil improvement on banks and slopes and in gullies where a permanent planting was desired, and it still provides these services to the South and its disturbed land. The plant has also crept into southern pine forests (the result of monocrop creation) and other weakened ecosystems. In this movement, kudzu replenishes, protects, and adds great biomass to depleted and disturbed soils.

Kudzu also provides nectar and pollen for insects and bees and shelter for a variety of animals.

Harvesting and Preparing

Harvest and prepare as usual. Due to its large size, harvesting kudzu root might require a backhoe or a very dedicated person with a shovel. Though there should be plenty of smaller roots to accomplish smaller yields. Collect the roots in the fall through early spring. To prepare, clean and shave off the outer bark, then cut into slices to dry. Once dry, it can be pulverized with mortar and pestle or a strong grinder for use as food or tea.

Dosage

High- to medium-dose, food-grade herbal.

Decoction: The TCM dosage is 10 to 20 grams of root, with a maximum of 60 grams. The TCM dosage of 3 to 12 grams of the flower is used for hangovers and excessive alcohol intake and the associated headache, fever, thirst, nausea, and vomiting.

Cautions and contraindications: Kudzu is a very safe, food-grade herb, though it can cause nausea and vomiting with a dose that is too high in individuals with weak digestion.

Herb and drug interactions: Use with caution with insulin and other antidiabetic medications, for there may be synergistic effects that may result in hypoglycemia. Also use caution taking the herb with any anticoagulant (heparin, warfarin or Coumadin), enoxaparin (Lovenox), and any antiplatelet medications (aspirin, dipyridamole or Persantine, and clopidogrel or Plavix) because kudzu has an antiplatelet effect.

References

K. O. Britton, D. Orr, and J. Sun, "Kudzu," in R. Van Driesche, et al., 2002, "Biological Control of Invasive Plants in the Eastern United States," USDA Forest Service Publication FHTET-2002-04, 413. www.invasive.org/eastern/biocontrol/25Kudzu.html (accessed March 8, 2010).

L. Wu, H. Qiao, Y. Li, et al., "Protective Roles of Puerarin and Danshensu on Acute Ischemic Myocardial Injury in Rats," *Phytomedicine* 14, no. 10 (October 2007): 652–58.

Ryan McGee, "Leafy, Invasive, Irritating Kudzu Could be NASCAR's Fuel of the Future," *ESPN The Magazine* (July 10, 2008).

F. M. Su, J. S. Chang, K. C. Wang, et al., "A Water Extract of *Pueraria lobata* Inhibited Cytotoxicity of Enterovirus 71 in a Human Foreskin Fibroblast Cell Line," *Kaohsiung Journal of Medical Science* 24, no. 10 (October 2008): 523–30.

Y. P. Hwang, C. Y. Choi, Y. C. Chung, et al., "Protective Effects of Puerarin on Carbon Tetrachloride-induced Hepatotoxicity," *Arch Pharm Res* 30, no. 10 (October 2007): 1309–17.

Y. J. Lin, Y. C. Hou, C. H. Lin, et al., "Puerariae Radix Isoflavones and Their Metabolites Inhibit Growth and Induce Apoptosis in Breast Cancer Cells," *Biochem Biophys Res Commun* 378, no. 4 (January 2009): 683–88.

Y. P. Hwang and H. G. Jeong, "Mechanism of Phytoestrogen Puerarin-mediated Cytoprotection Following Oxidative Injury: Estrogen Receptor-dependent Up-regulation of PI3K/Akt and HO-1," *Toxicol Appl Pharmacol* 233, no. 3 (December 15, 2008): 371–81.

H. Y. Zhang, Y. H. Liu, H. Q. Wang, et al., "Puerarin Protects PC12 Cells Against Beta-Amyloid-induced Cell Injury," *Cell Biol Int* 32, no. 10 (October 2008): 1230–37.

S. L. Connell, and S. H. Al-Hamdani, "Selected Physiological Responses of Kudzu to Different Chromium Concentrations," *Canadian Journal of Plant Science* 81, no. 1 (January 2001): 53–58.

Oriental Bittersweet
Celastrus orbiculatus

Common names: Oriental bittersweet, black-oil tree, climbing staff plant, intellect tree

Family: Staff-tree family (Celastraceae)

Related species: American bittersweet (*Celastrus scandens*), Indian bittersweet (*Celastrus paniculatus*)

States where Oriental bittersweet is considered invasive

The Plant

The bitterness many feel for *Celastrus* has forever been coupled with the undertones of sweetness the plant exudes. Its relentless taking over of the boundaries of disturbed landscapes and the way that it vigorously expands from there are balanced by the nurturing essence of the plant—its ability to heal these places and relax these boundaries. The sharp contrasts that are created from the cutting up of ecosystems must be softened by plants such as Oriental bittersweet, which create a barrier to further intrusion. This is how its medicine works for both ecosystems and people: it helps create a barrier from toxic influences and relaxes the stress and tensions of such disturbances.

While it does this, it begins to heal and lend support to disconnected nervous systems within our bodies, mirroring its aboveground growth habit of tenacious vines that weave and reconnect the gaps in an ecosytem's nerve junctions. For its warming beauty, we have placed it throughout homes in arrangements that stand on the mantel, therefore connecting us to nature. The sweetness of its medicine has been captured by people from its Far East native lands. It has been used to strengthen intellect and

244 ⅔ *Guide to Invasive Plants*

memory, relax the mind and body in times of pain, and to protect from toxic intrusions.

The traditional uses of Oriental bittersweet lend credence to its use in invasive pandemics. The medicine has long been used in the treatment of toxic influences and infectious disease—specifically, syphilis, which, like Lyme disease, is a spirochete-based infection. In addition, Oriental bittersweet has an affinity to the areas of the body where Lyme bacteria often inhabit, such as the joints, skin, and central nervous system. The plant helps the liver process toxins and has been an antidote for snakebites and opium poisoning, which are precursors to the modern-day toxins that so disturb humans.

Description

Oriental bittersweet is most notable for its round orange and yellow fruits that drape from leave-less trees in the winter and decorate flower arrangements during the holidays. Bittersweet is a climbing, twining shrub-vine that grows up to 60 feet into trees. Oblong, wavy, and slightly toothed leaves (2–5 inches long) alternate along the stem. In the spring emerge inconspicuous greenish yellow, five-petaled flowers in small clusters at the end of the branches. These form round green fruit that then become yellow and orange in the fall.

Collection and Habitat

After arriving to the United States from Far East Asia in the mid-1800s, Oriental bittersweet now occupies nearly the whole eastern half of North America. It is a widely expansive, vigorous vine that thrives on the boundaries of disturbed areas such as forest edges and gaps, fencerows and old fields, city lots, and in coastal areas.

Medicinal Uses

Parts used: Root, stem, leaf, and seed

TCM: Nan she teng (*C. orbiculatus*), root, stem, leaf; warm, neutral; associated with the liver and kidneys

1. Clears heat and toxins
2. Invigorates qi and blood
3. Dries dampness
4. Strengthens sinews and bones

Classically, bittersweet is used to treat paralysis, numbness of extremities, rheumatism, headache, toothache, abscesses, snakebite, anemia, beri-beri, carbuncles, colic, fever, gout, sores, and syphilis.

Ayurveda: Jyotishmati (*C. paniculatus*), leaf, seed, oil

Classically, bittersweet is used as a brain tonic to enhance intellect and memory (seed oil mixed with ghee) and to ease headache, depression, swooning, neurological disorders; urinary infections, sores, ulcers, wounds, scabies, eczema, swollen veins, intestinal parasites, rheumatism, lumbago, gout, paralysis, and pneumonia. The plant is used in Burma and the Philippines as an antidote to opium poisoning, and Indonesians use the leaves to treat dysentery.

Western botanical: Eclectics used the root bark and bark of *Celastrus scandens* to treat scrofula, syphilis, chronic liver disease, skin wounds, itching, burns, rheumatism, menstrual disorders, and painful breasts in nursing mothers.

Plant Chemistry

Acetoacetate, acetone, benzene, celaphanol A, celastrin, celastrol, epiafzelechin, epicatechin, n-butanol, paniculatin

Pharmacological Actions

Anticancer, analgesic, anodyne, antidotal, anti-inflammatory, aphrodisiac, diaphoretic, diuretic, ecbolic, emetic, emmenagogue, stimulant, tonic, antistress, antiaging, cognition facilitating

Scientific Studies

Anti-cancer: Numerous compounds found in *C. orbiculatus* have been deemed to possess antitumor and antioxidant abilities. In one study, two of the compounds, acetoacetate and n-butanol, inhibited the growth of tumors in mice. Other compounds of Oriental bittersweet have been said to partially or completely reverse resistance to multidrug-resistant cancer cells. An in vitro study using six compounds from *C. vulcanicola* showed an effectiveness that was similar to, or higher than, the classical P-Glycoprotein reversal agent verapamil for the reversal of resistance to daunomycin and vinblastine. Also in in vitro studies, the fruit showed promising cytotoxicity against both human melanoma and human cervical carcinoma.

Antioxidant: In India and other parts of the world, *C. paniculatus* seed oil has been studied extensively for its antioxidant, anti-inflammatory, tranquillizing, cognitive enhancing, anticonvulsant, and central nervous system tonic properties. In one study, the *C. paniculatus* extract showed a dose-dependent free radical scavenging capacity and a protective effect on DNA cleavage.

Antimicrobial: The alcoholic emulsion of *C. paniculatus* seed oil has shown antibacterial activity against gram-positive and gram-negative bacteria.

Other Uses

Some Native Americans are said to have eaten the tender twigs and sweet inner bark, and in the East, in the spring, the young leaves are eaten boiled. Too much is toxic, however, and has been known to have poisoned horses.

Ecological Importance

No phytoremediation reports were found for Oriental bittersweet. Though, by virtue of its presence in often-polluted roadsides, parking lots, and river edges, this plant's natural attenuation abilities will be validated. This viny plant grows to protect fractured landscapes and acts as a tonic to disturbed ecological nervous systems.

The berries are also an important winter food source for birds.

Harvesting and Preparing

Harvest and prepare as described.

Dosage

Low- to medium-dose botanical. Many consider the plant to be toxic, especially the berry.

Decoction: The TCM dosage is 15 to 30 grams of stem and leaf, taken internally. The fresh leaves are used externally. The ayurvedic dosage is 1 to 2 grams of the leaf and seed and 5 to 15 drops of the seed oil.

Cautions and contraindications: Oil is emetic and toxic in large doses, the bark extracts are thought to be cardioactive, and the fruit is considered toxic, so many sources point to the potential risk of ingesting bittersweet. Advice from a herbal health care provider familiar with this plant is recommended when using this herb.

Herb and drug interactions: None known.

References

J. Zhang, Y. M. Xu, W. M. Wang, et al., "Experimental Study on Anti-tumor Effect of Extractive from *Celastrus orbiculatus In Vivo*," *Zhongguo Zhong Yao Za Zhi* 31, no. 18 (2006): 1514–16.

M. Gattu, K. L. Boss, A. V. Terry Jr., et al., "Reversal of Scopolamine-induced Deficits in Navigational Memory Performance by the Seed Oil of *Celastrus paniculatus*," *Pharmacol Biochem Behav* 57, no. 4 (1997): 793–99.

F. Ahmad, R. A. Khan, and S. Rasheed, "Preliminary Screening of Methanolic Extracts of *Celastrus paniculatus* and *Tecomella undulata* for Analgesic and Anti-inflammatory Activities," *J Ethnopharmacol* 42, no. 3 (1994): 193–98.

H. Z. Jin, B. Y. Hwang, H. S. Kim, et al., "Anti-inflammatory Constituents of *Celastrus orbiculatus* Inhibit the NF-kappa B Activation and NO Production," *J Nat Prod* 65, no. 1 (2002): 89–91.

S. E. Kim, Y. H. Kim, J. J. Lee, et al., "A New Sesquiterpene Ester from *Celastrus orbiculatus* Reversing Multidrug Resistance in Cancer Cells," *J Nat Prod* 61, no. 1 (1998): 108–11.

A. Russo, A. A. Izzo, V. Cardile, et al., "Indian Medicinal Plants as Antiradicals and DNA Cleavage Protectors," *Phytomedicine* 8, no. 2 (2001): 125–32.

J. Xu, Y. Q. Guo, X. Li, et al., "Cytotoxic Beta-dihydroagarofuran Sesquiterpenoids from the Fruits of *Celastrus orbiculatus*," *Z Naturforsch* [C]. 63, no. 7–8 (2008): 515–18.

D. Torres-Romero, F. Muñoz-Martínez, I. A. Jiménez, et al., "Novel Dihydro-beta-agarofuran Sesquiterpenes as Potent Modulators of Human P-glycoprotein Dependent Multidrug Resistance," *Org Biomol Chem* 7, no. 24 (2009): 5166–72.

States where plantain is considered invasive

Plantain
Plantago spp.

Common names: broadleaf plantain, common plantain, rippleseed plantain (*P. major*), lance-leafed plantain, narrowleaf plantain (*P. lanceolata*)

Family: Plantain family (Plantaginaceae)

Related species: There are 34 *Plantago* species growing the world over with many unique varieties that have each adapted to distinct ecologies. Asian plantain (*Plantago asiatica*), Psyllium, sand plantain (*P. psyllium*), desert indianwheat (*P. ovate*), and woolly plantain (*P. patagonica*).

Praying, I took a piece of plantain leaf in my mouth. The bite was already quite inflamed with a deep, black and blue mark where the tick had embedded itself. I chewed the leaf, macerating it between my teeth, sucking the juice from it—a bittersweet medicine that I swallowed. As the leaf became paste, I rubbed it on the bite and covered it with a bandage. The medicine is like a lance that penetrates and opens the wound to draw out the poison. Humans have probably used plantain for such a purpose for hundreds of thousands of years. This connection to the plant is deeply ingrained in my genetics—and within all of ours. It is a gift of nature that abounds for all of us, and I give thanks for this blessing.

The Plant

I highlight plantain not necessarily because many consider it noxious, but because of its abundance throughout the world, inhabiting backyards, roadsides, parking lots, and almost anywhere else. Indeed, a variety of plantain species have adapted to most locales that have been disturbed by humans. This plant has traveled the world over with Europeans and the people of North America and Asia. *Plantago major* had been named English-Mans Foot by some Native Americans because it seemed to fall from settlers' boots. For more than two thousand years, traditional Chinese medicine has documented its own plantain species for use, and the ancients highly esteemed this plant. It removes toxins from the body and addresses infections, boils, bites, stings, tumors, inflammation, and pain. Because of its tannins, when taken internally it removes the toxic affliction from tissues, and expels it, primarily through urination. Through its draining and cooling powers, the inflamed tissues and pain are resolved. It is an esteemed folk remedy for various cancers and infectious diseases, and modern science has finally caught up to what has already been held as true by traditions throughout the world for thousands of years.

Plantain tempers the fiery arrogance of a world gone mad.

Description

Plantain is a perennial weed with basal leaves that are often seen throughout the year. *P. major* has wavy, ribbed, and broad oval-shaped leaves, and *P. lanceolata* has thin, lance-shaped leaves with three ribs. In the late spring through fall both species produce tiny, whitish flowers in a cylindrical head (0.5–1.5 inches long) upon a stalk 6 to 12 inches tall that has emerged from the basal leaves. After flowering, minuscule black seeds ripen.

Collection and Habitat

Plantain species are found all over the world, with *P. major* and *P. lanceolata* present in every region of North America except for the far northern arctic region (where very few grow). The Latin name for plantain comes from *planta,* "sole of the foot," and this accurately describes the way in which this plant travels around with humans. Plantain is found along roadsides, trails, sidewalks, and almost anywhere else the human foot causes disturbance and extreme conditions, and it survives where few other plants will.

The plant needs us, and we need it. We will never be separated.

Medicinal Uses

Parts used: Root, leaf, flower spikes, and seed

TCM: Che qian zi (*Plantago asiatica semen*), "before the cart seeds," seeds; che qian cao, herb; sweet, cold; associated with the bladder, kidneys, liver, and lungs

> 1. Promotes urination and clears damp heat
> 2. Resolves dampness and stops diarrhea
> 3. Clears liver heat and benefits the eyes
> 4. Clears phlegm from the lungs

Classically, plantain is used to treat burning, painful, difficult urination; diarrhea; eye redness, swelling, ulcers, sores, pain; headaches, dizziness, blurred vision, summer heat, rheumatism, cough, infectious diseases of repiratory, urinary, and digestive systems.

Ayurveda: Isabgola, Ashvagola (*P. ovate*), seed, husk

Used as in TCM, but ayurveda additionally uses the seeds and husks for chronic constipation.

Western botanical: In *Handbook of Medicinal Herbs,* James Duke presents the numerous plantain remedies from throughout the world for treatment of asthma, bronchitis, bruises, boils, cancer, cold, cough, convulsions, diarrhea, dropsy, dysentery, earache, enuresis, epilepsy, epistaxis, fever, gonorrhea, gout, headache, hemoptysis, hemorrhoids, hepatitis, herpes, jaundice, kidney stones, lunacy, malaria, piles, otitis, puerperium, renitis, ringworm, shingles, sore throat, splenitis, stangury, syphilis, thrush, toothache, and ulcers.

Plantain has been used for various cancers throughout the body and is prepared as medicine throughout the world (most notably in Latin America, where it is known as *el llanten*). The native peoples of the Americas all made good use of various *Plantago* species—in very much the same ways that other cultures have used it.

The astringency of plantain draws out inflammation and infection, and, as a diuretic, it sends this toxic buildup out through the bladder. The plant is active against a broad range of organisms, inhibiting bacteria, viruses, parasites, fungi, and cancer with its antimicrobial nature. With potent antioxidants, micronutrients, and sulfur compounds, it appears that plantain can assist in the chelating of heavy metals that are stored in the body's tissues. With traditional use and modern studies showing plantain to be a detoxifier, blood cleanser, liver protectant, and immune stimulator, and with its mineral content to bind to heavy metals, the prospect is promising for employing plantain. One aspect of this can be seen in a study that showed plantain as inhibiting the absorption of some medications, including lithium, by binding to this chemical and excreting it from the body.

Plant Chemistry

Acetoside, aucubin, allantoin, adenine, apigenin, asperuloside, baicalein, baicalin, benzoic acid, caffeic acid, catalpol, chlorogenic acid, choline, cinnamic acid, d-glucose, emulsin, ferulic acid, fructose, fumaric acid, geniposidic acid, gentisic acid, glucoraphenine, hispidulin, indicain, invertin, lignoceric acid, loliolid, luteolin, neo-chlorogenic acid, nepetin, oleanolic acid, p-coumaric acid, plantagonine, plantagoside, plantenolic acid, plantease, potassium salts, sacchorose, salicyclic acid, scutellarin, sorbitol, sulforaphene, sitosterol, syringin, tyrosol, ursolic acid, and vanillic acid

Pharmacological Actions

Alterative, astringent, anticancer, antimicrobial, demulcent, deobstuent, diuretic, expectorant, refrigerant, styptic, anodyne, anticandidal, immunomodulatory

Scientific Studies

Researchers have isolated the potent compounds found in Plantain species, validating the long-traditional uses of the plant to enhance immunity and to treat various cancers and infectious disease.

Anticancer: In one study, under the guidelines recommended by the National Cancer Institute for investigating cytotoxic activity, extracts from *Plantago* species were found to inhibit the proliferation of human cancer cell lines. This was due to the flavonoids—most notably luteolin—present in plantain. Another study with rats found that if a plantain polyphenolic complex is included in their diet while they are subjected to drug-induced toxic liver damage and tumor creation, they normalized the biochemical parameters and also decreased the tumor size from 87.5 percent to 33.3 percent. In this case, the compound plantastine was found to be a main contributor to the plant's ability to prevent and reverse cancer. Other studies have put forth evidence in support of considering plantain for the treatment of various cancers and for treating artificially created tumors. An abstract report from Taiwan appears to sum up its potential in cancer treatment.

> In this study, we investigated the antiviral, cytotoxic and immunomodulatory activities of hot water extracts of [*Plantago major, P. asiatica*] *in vitro* on a series of viruses, namely herpesviruses, adenoviruses, and on various human leukemia, lymphoma and carcinoma cells. Results showed that hot water extract of *P. asiatica* possessed significant inhibitory activity on the proliferation of lymphoma and carcinoma (bladder, bone, cervix, kidney, lungs, and stomach) cells and on viral infection. *P. major* and *P. asiatica* both exhibited dual effects of immunodulatory activity, enhancing lymphocyte proliferation and secretion of interferon-gamma at low concentrations, but inhibiting this effect at high concentration. The present study concludes that hot water extracts of *P. major* and *P. asiatica* possess a broad spectrum of antileukemia, anticarcinoma and antiviral activities, as well as activities which modulate cell-mediated immunity.[23]

An in vivo study on mice also found that *P. major* had an inhibitory effect on Ehrlich ascites tumors.

Antimicrobial: Additional in vitro studies have found *Plantago* extracts inhibiting the growth of *Bacillus subtilis, Bacillus dysentery, E. coli, Streptococcus pneumonia,* and *Giardia*. Further, it contains ferulic acid and caffeic acid, which have demonstrated in vitro to act against herpes simplex virus 2 (HSV-2) and adenovirus (ADV-3).

Respiratory: In a clinical trial of plantain in the treatment of chronic bronchitis, twenty-five patients, some with spastic character, were given the plant for twenty-five to thirty days. Rapid and successful treatment was obtained in 80 percent of the subjects, with overall improvement in respiration and no toxic side effects.

Other Uses

The young leaves are tasty raw, in salads, and the older leaves can be cooked as greens or added to soup. Avoid the hairy varieties of *Plantago*. I usually stick to young leaves of *P. major* or *lanceolata*. Young flowering stalks are reported by François Couplan to be "delicious simply sautéed in butter."[24] Though a relative of the bulking agent psyllium, common plantain seed is not generally considered edible; Dr. Duke presents his esteemed version of poor-man's branflakes (or Metamucil), with milk and sugar, as the seeds have a curdling effect on milk. The seeds do make good chicken food, and the herb is agreeable fodder to some livestock (sheep, goats, pigs) and has also long been given to farm animals as medicine.

Ecological Importance

The essence of plantain's importance in the ecosystem can be summed up in the same terms as the medicine it provides to warm-blooded creatures. It leaches toxic influences from the body and does this with the landscape as well. *Plantago* grows along continually trodden paths, striving to bring life to desolate roadsides, trails, gravel pits, and anywhere the cart has passed—areas that are continually irritated by the ongoing wear of foot and wheel, the exposed areas where the sun burns many other plants. Plantain survives in such places and does its best to stabilize the bombardment of "white man's footsteps."

One study found plantain growing comfortably in an oil-polluted environment, and it was a predominate species in the gravelly, contaminated soils. Because of its tolerance to NO_2 and SO_2, the plant was also used in a remediation study; the study also addressed whether either of these pollutants impact the plant's ability to uptake them. In another study, plantain was able to phytostabilize copper in contaminated soil and thereby to inhibit its further spread into the surrounding environment. With time, we will find further evidence of this plant's wonders in detoxifying the damaged landscape.

As a wildlife food, plantain is sold as birdseed. Little birds especially like it. It is also a nectar source for bees, ants, and other insects that make use of nectar and pollen.

Harvesting and Preparing

Harvest and prepare as described. The herb can be prepared externally as a poultice, as a wash for wounds, or as an eyewash for irritations. It can also be used in the bathtub for soaking.

Dosage

Medium- to high-dose botanical.

Decoction: The TCM dosage is 9 to 15 grams of the seed.
Tincture: 20 to 60 drops

Cautions and contraindications: It is a very safe herb.

Herb and drug interactions: Use caution with diuretic medications. Plantain could limit absorption of other medications as well, including those for diabetes and their uptake of carbohydrates.

References

M. Ozaslan, I. D. Karagoz, I. H. Kilic, et al., "Effect of *Plantago major* Sap on Ehrlich Ascites Tumours in Mice," *African Journal of Biotechnology* 8, no. 6 (March 20, 2009): 955–59.

M. Gálvez, C. Martín-Cordero, M. López-Lázaro, et al., "Cytotoxic Effect of *Plantago* spp. on Cancer Cell Lines," *Journal of Ethnopharmacology* 88, no. 2–3 (October 2003): 125–30.

M. Matev, I Angelova, A. Koĭchev, et al., "Clinical Trial of a *Plantago major* Preparation in the Treatment of Chronic Bronchitis," *Vutr Boles* 21, no. 2 (1982): 133–37.

L. C. Chiang, W. Chiang, M. Y. Chang, et al., "In Vitro Cytotoxic, Antiviral and Immunomodulatory Effects of *Plantago major* and *Plantago asiatica*," *American Journal of Chinese Medicine* 31, no. 2 (2003): 225–34.

J. Yoon, X. Cao, X. Zhou, et al., "Accumulation of Pb, Cu, and Zn in Native Plants Growing on a Contaminated Florida Site," *Science of the Total Environment* 368 (2006): 456–64.

M. Ponce-Macotela, I. Navarro-Alegría, M. N. Martínez-Gordillo, et al., "*In Vitro* Effect against Giardia of 14 Plant Extracts," *Rev Invest Clin* 46, no. 5 (1994): 343–47.

L. C. Chiang, W. Chiang, M. Y. Chang, et al., "Antiviral Activity of *Plantago major* Extracts and Related Compounds *In Vitro*," *Antiviral Res* 55, no. 1 (2002): 53–62.

M. Ozaslan, I. Didem Karagöz, M. E. Kalender, et al., "*In Vivo* Antitumoral Effect of *Plantago major* L. Extract on Balb/C Mouse with Ehrlich Ascites Tumor," *Am J Chin Med* 35, no. 5 (2007): 841–51.

R. Velasco-Lezama, R. Tapia-Aguilar, R. Román-Ramos, et al., "Effect of *Plantago major* on Cell Proliferation *In Vitro*," *J Ethnopharmacol* 103, no. 1 (2006): 36–42.

R. Gomez-Flores, C. L. Calderon, L. W. Scheibel, et al., "Immunoenhancing Properties of *Plantago major* Leaf Extract," *Phytother Res* 14, no. 8 (2000): 617–22.

L. C. Chiang, L. T. Ng, W. Chiang, et al., "Immunomodulatory Activities of Flavonoids, Monoterpenoids, Triterpenoids, Iridoid Glycosides and Phenolic Compounds of *Plantago* Species," *Planta Med* 69, no. 7 (2003): 600–604.

A. B. Samuelsen, "The Traditional Uses, Chemical Constituents and Biological Activities of *Plantago major* L. A Review," *J Ethnopharmacol* 71, no. 1–2 (2000): 1–21.

H. J. Taylor and J. N. Bell, "Tolerance to SO2, NO2 and Their Mixture in *Plantago major* L. Populations," *Environ Pollut* 76, no. 1 (1992): 19–24.

F. B. Holetz, G. L. Pessini, N. R. Sanches, et al., "Screening of Some Plants Used in the Brazilian Folk Medicine for the Treatment of Infectious Diseases," *Mem Inst Oswaldo Cruz* 97, no. 7 (2002): 1027–31.

Purple Loosestrife
Lythrum salicaria

Common names: loosestrife, lythrum, purple willow herb, spiked loosestrife, blooming Sally

Family: Loosestrife family (Lythraceae)

Related species: *Lythrum alatum, L. vulgaris, L. album, L. anceps, L. lanceloatum.* However, a number of loosestrifes—including yellow loosestrife—comes from a different family.

States where purple loostrife is considered invasive

The Plant

Purple loosestrife is a handsome roadside companion throughout the United States, stretching for miles in the median of the highway and out into the surrounding lowlands and wetlands. It is another valuable medicine that has hitched a ride into this country from Europe and Asia, but even in the 1800s this plant was so plentiful that some botanists thought it was native. It has spread deep into our lands, and, according to some reports, forty-five million dollars are spent every year in unsuccessful attempts to rid the environment of it. Purple loosestrife is the poster plant for invasive ideology and receives notable attention for its widespread presence. Although originally brought to this country because it was known as a beautiful flowering ornamental, many people now regard the presence of purple loosestrife as "the purple plague." The attack on purple loosestrife continues, even without scientific evidence that it proves harmful to its environments. In fact, purple loosestrife is rehabilitating the wetlands with its ability to absorb from the water the excess nitrogen and phosphorus that results from agricultural fertilizer and pesticide runoff, and it also helps prevent soil erosion in disturbed wetland areas.

The equating of invasive and bad blinds us to the potentials of invasive plants and the potential environmental good these plants provide. Todd Hardie of Honey Gardens in Vermont has experienced this negative mentality from individuals and explains the importance of this plant to beekeepers.

> Over the years, this has been one of the most important nectar and pollen plants for our bees. We have had bee yards that were in the midst of 100–1,000+ acre fields of purple loose strife. While I understand that it is considered "invasive," there is another side to the story. The nectar and pollen from this plant gives life to pollinating insects and helps to sustain them through these challenging times for the honey bees. . . . We once had a lot of pure purple loosestrife honey, and I was so excited to share this with our markets that I got a purple label and noted the floral source on the top of each jar, "purple loosestrife." Soon after the honey reached the market, the calls started coming in, criticizing us for being involved with this honey. Most every day, someone would call and get mad at me. I explained that we were not spreading the flowers, and that it was against the law to do so, but that the bees were working on the flowers, and this is what they made. I could not even get to the part of the story that the nectar and pollen were supporting the pollinating insects.[25]

In addition to the flowers imparting a subtle flavor to the honey, they are also infusing it with medicine the plant provides. In the honey-making process, the bees also receive healing from the pollination dance they do and are infused with the plant's medicine as well, which is sorely needed, considering the present collapse of many bee colonies. Research in Australia has found that honey made from bees that feed on tea tree (*Leptospermum*) has potent qualities that can treat antibiotic-resistant infections. Purple loosestrife has been found to possess powerful antimicrobial properties of its own and has been used for thousands of years by herbalists to treat infections of the respiratory and digestive systems. Clearly purple loosestrife is of enormous value for both bees and people.

> *The name* loosestrife *comes from a long-ago time and means "dissolving strife." It refers to the use of the plant to "quieten savage beasts."*[26]

Description

Purple loosestrife is a perennial plant that is often noticed when it establishes vast purple stands that carpet wetlands. A single rootstock can cover many square feet and send up thirty to fifty erect stems, each growing up to 8 feet tall. Narrow leaves with smooth edges whorl around the angular stem. Starting in the summer and into fall, *Lythrum salicaria* produces lovely purple and/or pink flowers with five to seven petals, arranged along the top of the stem and forming long purple spikes. The flowers open from the bottom up along the spike, and eventually in the late fall, a loosestrife plant produces seeds, sometimes as many as two million seeds in one season.

Collection and Habitat

Originally from Europe and Asia, purple loosestrife arrived in North America with European colonization and was so common by the early 1800s it was believed by many botanists to be a native plant. The plant is now found throughout the Canadian provinces and lower forty-eight states, with the possible exception of Florida, inhabiting wetland areas such as bogs, marshes, floodplains, rivers, and lake edges. Numerous factors have contributed to purple loosestrife spreading throughout the land and waterways, mostly with the help of humans. In early colonial days, the plant was admired by horticulturists and was planted in many gardens. Bedding and feed for livestock, and even sheep wool, most likely carried the seed. The plant also caught rides in the ballasts of European ships and spread through the canals and waterways that were later dug out and which connected vast networks of wetlands and rivers. Beekeepers had a special affinitiy for purple loosestrife in honey making, for its abundant blooms and ability to flourish in bogs and marshes. In fact, honey manufacturing accounts for much of the plants movement into the Midwest and Far West.

Medicinal Uses

Parts used: The whole flowering plant

TCM: Qia qu cai (*Herba Lythrum*); cold, bitter, astringent; associated with liver, lung, large intestine. Classically, loosestrife is used for bacillary dysentery.

Ayurveda: No known use.

Western botanical: European folk tradition uses purple loosestrife for diarrhea, dysentery, ophthalmia, piles, ulcers, skin erruptions, typhus, wounds, stops bleeding, sore throats, gonorrhea, cholera, liver disease, and leucorrhea. Related species have been used in the treatment of cancer. According to Maud Grieve in *A Modern Herbal,* purple loosestrife is "superior to Eyebright for preserving the sight and curing sore eyes . . . even in blindness if the crystalline humour is not destroyed."[27] The herb has also been successfully used in an epidemic of dysentery. Rosemary Gladstar recounts purple loosestrife's use "for bronchial and lung infections, as used in European medicine. It's a great herb for the immune system, especially blended with its wetland neighbors, Boneset and Joe Pye Weed."[28]

Dioscorides (first century) is credited as the earliest authority on purple loosestrife. He reported,

> The herb is tart and strong in taste, of an astringent and refrigerant nature, good for stanching both outward and inward bleeding; sap extracted from the leaves and drunk stops blood-spitting and dysentery, and sour wine in which the leaves have been boiled when taken internally will have the same effect; and if the plant is set afire it gives off a pungent vapor and smoke that drives away serpents; and flies cannot stay in a room where this smoke is.[29]

Today, purple loosestrife is regarded as an undervalued medicinal that has potential for cleaning pandemic infections and leaching toxins.

Plant Chemistry

Anthocyanins, B-sitosterol, butyl-phthalate, cholin, chlorogenic acid, cyaniding-3-monogalactoside, dibutyl-phthalide, ellagic acid, glycoside salcarin, loliolide, malvidin-3,5-diglucoside, malvin, narcissi, oleanolic acid, ursolic acid, orientin, isoorientin, isovitexin, p-coumaric acid, pectin, salicarin, tannins, vescalagin, vitexin ellagitannins, pedunculagin, vanoleic acid dilactone, and galloylglucose

Pharmacological Actions

Antimicrobial, antioxidant, antitumoral, anticancer, anti-inflammatory, hypoglycemic, astringent, demulcent, expectorant, hepatoprotective, styptic, anthelmintic, emmenagogue, fungicidal, sedative, and vulnerary

Scientific Studies

Antimicrobial: Purple loosestrife extracts have found to be effective in vitro in inhibiting the growth of *Candida albicans, Staphylococcus aureus, Proteus mirabilis, Microccocus luteus, E. coli,* and the phytopathogenic fungus *Cladosporium cucumerinum.* This has been attributed to the antifungal agents oleanolic and ursolic acid and the antibacterial compound vescalagin.

Antioxidant: Purple loosestrife extract has been found to have antioxidant effect due to the flavonoids isovitexin and isoorientin. The extract was found to be active in both in vitro antioxidant studies and in vivo pharmacological activity tests and was confirmed for it is free radical–scavenging ability and for chelating excess iron.

Anti-inflammatory: The flavonoids isovitexin and isoorientin have also displayed anti-inflammatory actions.

Other Uses

The young shoots of purple loosestrife have been used as an emergency food source.

American beekeepers have recognized the importance of purple loosestrife in the production of honey: It is recommended by honey plant enthusiasts for its abundant flowers and for being "the one plant which can be planted with confidence in boggy spots."[30] Purple loosestrife has infused many honey jars and has sweetened a great deal of tea.

Loosestrife is also used in tanning leather, and one folk use involves burning it as incense to repel flies, gnats, and snakes.

Ecological Importance

This wetland plant helps cleanse polluted waters by absorbing nitrogen and phosphorus and tolerating high levels of lead and other toxins that are prevalent in wetland ecosystems. In a study of this most important wetland cleaner, purple loosestrife was shown to exhibit greater growth than most of the other wetland plants and was efficient at improving nutrient uptake and balancing soil microbes through its ability to aerate the soil. In this, it acts as an antioxidant for the wetland ecosystem. It also helps to prevent soil erosion.

Purple loosestrife makes a good substitute for the endangered plant eyebright, which has been overharvested in the wild for medicine. With its abundant flowers, it provides nectar for insects and bees.

Harvesting and Preparing

When the plant is in bloom, harvest and prepare as usual. Externally, the fresh or dried herb is prepared as an ointment and wash for wounds, ulcers, sores, scabs, and irritated eyes.

Dosage

Medium-dose botanical.

Decoction: 3 to 6 grams

Tincture: 20 to 30 drops

Cautions and contraindications: None known.

Herb and drug interactions: None known.

References

J. P. Rauha, S. Remes, M. Heinonen, et al., "Antimicrobial Effects of Finnish Plant Extracts Containing Flavonoids and Other Phenolic Compounds," *International Journal of Food Microbiology* 56, no. 1 (May 25, 2000): 3–12.

Z. Tunalier, M. Koşar, E. Küpeli, et al., "Antioxidant, Anti-inflammatory, Anti-nociceptive Activities and Composition of *Lythrum salicaria* L. Extracts," *Journal of Ethnopharmacology* 110, no. 3 (April 4, 2007): 539–47.

J. P. Rauha, J. L. Wolfender, J. P. Salminen, et al., "Characterization of the Polyphenolic Composition of Purple Loosestrife (*Lythrum salicaria*)," *Z Naturforsch [C]* 56, no. 1–2 (January–February 2001): 13–20.

H. Becker, J. M. Scher, J. B. Speakman, et al., "Bioactivity Guided Isolation of Antimicrobial Compounds from *Lythrum salicaria*," *Fitoterapia* 76, no. 6 (September 2005): 580–84.

Daniel Q. Thompson, R. L. Stuckey, and E. B. Thompson, "Spread, Impact, and Control of Purple Loosestrife (*Lythrum salicaria*) in North American Wetlands," United States Department of the Interior—Fish and Wildlife Service, www.npwrc.usgs.gov/resource/plants/loosstrf/index.htm (accessed December 16, 2008).

J. L. Uveges, A. L. Corbett, and T. K. Mal, "Effects of Lead Contamination on the Growth of *Lythrum salicaria* (Purple Loosestrife)," *Environmental Pollution* 120, no. 2 (2002): 319–23.

J. Villaseñor Camacho, A. De Lucas Martínez, R. Gómez Gómez, et al., "A Comparative Study of Five Horizontal Subsurface Flow Constructed Wetlands Using Different Plant Species for Domestic Wastewater Treatment," *Environmental Technology* 28, no. 12 (December 2007): 1333–43.

Dana Joel Gattuso, "Invasive Species: Animal, Vegetable, or Political?" *National Policy Analysis,* no. 544 (August 2006).

States where common reed
is considered invasive

Reed
Phragmites australis

Common names: common reed, giant reed,
ditch reed, phragmites

Family: Grass family (Poaceae)

Related species: There are at least twenty
genetic lineages of phragmites, and eleven are
considered native to North America, many
difficult to tell apart. Though some say there
is only one other plant in the same genus
Phragmites: tall reed (*P. karka*).

The Plant

I was fortunate to have common reed introduced to me early on in my
quest to know the most important invasive plants. *Phragmites* is one of
the most widespread wetland plants in the world and receives bad press
for creating vast, imposing stands. I soon remembered its common usage
in Chinese herbal medicine (as Lu gen) for its heat-clearing, detoxifying,
anti-inflammatory properties in the upper (lung), middle (stomach), and
lower (bladder) parts of the body. What's more, it provides a remedy in
cases of seafood poisoning.

There is a reason it can be found growing next to the catch.

As I dived into the nature of this plant and wondered why it was growing where it does, I did not realize the buried treasure I would find just
below the surface of the water. *Phragmites* could very well be the most
important naturally dispersed, bioremediating plant species in the world.
As we have discovered, common reed displays the ability, under scientific
scrutiny, to clean sewage wastewater effectively, removing fifteen heavy
metals (Zn, Ph, P, Pb, Cd, Cr, Ni, Mn, Cu, Zn, Se, Na, Th, U, Cs-137)
and at least eleven common toxic pollutants (herbicides, petroleum, TNT,

DDT, PCBs, xenobiotics, chlorobenzene, phenols, sulfide, acid orange 7, polycyclic aromatic hydrocarbons). It is no wonder we find this plant growing in abundance along roadsides, drainage ditches, floodplains, and wetlands. *Phragmites* serves as the chief water filter and detoxifier throughout the earth's cleansing wetland systems. Common reed has been used for treatment of mine wastes; explosives; and agricultural, industrial, and municipal wastewaters throughout the world.

Overall, phragmites is perfect for those hot and toxic places, both inside the body and in the environment. Common reed has been traditionally used for various cancers, leukemia, fever, infections, foul sores, and is a modern-day ecological savior cleaning nearly every imaginable heavy metal, toxic chemical, explosive, pesticide, and waste by-product.

It's interesting to note the common threads between the human and ecological healing abilities of phragmites:

cancer and heavy metals,
leukemia and toxic chemicals,
fever and explosives,
infection and pesticides,
foul sores and waste by-products.

Description

Phragmites is a giant grass species, growing 10 to 20 feet (3–6 meters) tall. Its leaves emerging from the stalk are similar in appearance to corn, each 50 to 100 inches (20–40 cm) long, and about an inch wide (1–4 cm). Common reed can be easily distinguished by the fluffy plume of flowers that appear in the summer and wave high above the stems, oftentimes standing until the next growing season.

Collection and Habitat

Common reed is naturalized throughout the world, including the continental United States and southern Canada. Its native range is quite obscure—it has been floating on water and in the air, spreading for quite some time now, with some archaeological records showing *P. australis* present on some sites for almost three thousand years. *Phragmites* is common along roadsides in drainage ditches and can also be found in wet areas such as marshes, floodplains, water edges, and wet meadows while also tolerating brackish water.

Medicinal Uses

Parts used: Roots, stems, leaves, flowers, and seeds

TCM: Lu Gen (*Rhizoma Phragmitis*), "reed rhizome"; sweet, cold; associated with the lungs, stomach, bladder

1. Clears heat, generates fluids
2. Clears heat from the lungs
3. Clears heat from the stomach
4. Clears heat from the bladder, promotes urination
5. Vents rashes
6. Relieves food poisoning

Classically, common reed is used to treat fevers, irritability, sore throat, strong thirst, sunstroke, nausea, vomiting, cough with thick phlegm, bronchitis, lung abscesses with blood and pus, difficult urination, arthritis, early-stage measles, cholera, foul sores, toothache, diabetes, leukemia, breast cancer, and fish, crab, and seafood poisoning.

Ayurveda: Nala* (*Phragmites karka* or *Arundo donax*), Great Reed

Classically, Nala is used internally and externally to treat herpes, fever, skin inflammation and redness, internal bleeding, and urinary bladder infection.

Western botanical: Folk remedy for abscesses, arthritis, bronchitis, cancer, cholera, cough, diabetes, dropsy, dysuria, fever, flux, gout, hematuria, hemorrhage, hiccup, jaundice, leukemia, lung, nausea, rheumatism, sores, stomach, thirst, and typhoid. The juice of the plant is said to soothe insect bites. Native Americans used reed to treat diarrhea and stomach troubles, pneumonia, and to loosen phlegm and soothe lung pain.

Plant Chemistry

Arabinose, ascorbic acid, ash, asparagines, B-amyrin, beta-carotene, caffeic acid, calcium, chrysoeriol, coixol, coniferaldehyde, D-galactose, dioxane lihnin, ferulic acid, furfural, gentisic acid, glucose, guaiacyl, isoquercitrin, luteolin, p-benzoquinone, p-coumaric acid, p-hydroxybenzaldehyde, phosphorus, rutin, saponin, silicic acid, sucrose, syringaldehyde, syringyl, taraxerol, taraxerone, tocopherols, tricin, vanillic acid, wax, xylos

---

*According to *Indian Herbal Remedies,* although *Phragmites karka* is the accepted source of Nala in ayurvedic medicine, *Arundo donax* is also recognized.

Pharmacological Actions

Antimicrobial, gastrointestinal, anti-inflammatory, febrifuge, central nervous system suppressant, diuretic, diaphoretic, antidotal, antiemetic, antipyretic

Scientific Studies

Antimicrobial: *Phragmites* has been studied in vitro, proving effects against B-hemolytic *Streptococcus*. It has been studied within a formula in the successful treatment of pulmonary abscesses.

Antiaging: Researchers in China have found that a polysaccharide in *Phragmites* resists the atrophy of the thymus, spleen, and brain tissues of aging mice.

Other Uses

Native Americans used the rootstock by filtering water through the pounded root mass and collecting its starches that settle out. The young shoots can be eaten raw or steamed as well. Before flowering, the stem is rich in sugars and can be dried and powdered to add to flours as sweetener. The sap, which comes from the stalk, turns into a resin candy ball. The nutritious seeds can also be consumed and "are rich in protein, fairly easy to harvest and are quite tasty,"[31] according to Rosemary Gladstar. The younger plants can be given as fodder.

The stems have been made into mats to cover roofs and greenhouses, drying mats, baskets, flutes, arrow shafts, water pipes, and rope. Because of their high cellulose content and long fibers, they have also been used in the manufacture of pulp for paper and rayon.

The Iroquois used a decoction of common reed rootstocks and bottle brush grass ceremonially as a medicine for soaking corn seeds before planting.

Common reed has been used in Europe as a biomass source, and with the abundance of this plant and with enhanced harvesting methods, it could serve power needs for communities.

Ecological Importance

Phragmites is a valuable phytoremediator used in water filtration systems for homes, municipalities, industries, and government agencies all over the world. It is also considered an ideal species to phytostabilize and rehabilitate severely mined sites that have extreme pH, high salinity, and phytotoxins. Besides addressing the contaminants, phragmites was found to help accumulate soil organic matter with its loading biomass in order to re-create the natural habitat. Compared to other pioneer species that were estimated to take about one hundred twenty years of continuous growth to increase the organic matter at natural sites, phragmites was said to take thirty years. Because of its high capability to accumulate trace metals in the roots, phragmites is considered a good ecological monitor for contamination in lakes, waterways, and soils.

The dense stands of reed in wetlands provide an abundance of food for wildlife. The seeds feed birds, and the decaying plant matter enters estuarine food webs and provides for other organisms in marshes, in much the same way native cordgrass does.

Harvesting and Preparing

Harvest the roots in the fall to early spring and dry and prepare as described. Its fresh form is said to have a stronger function in clearing heat and promoting bodily fluids in cases of dehydration. For external use, it can be prepared as a poultice and wash.

Dosage

Medium-dose botanical.

Decoction: The TCM dosage is 10 to 30 grams of dried and 15 to 60 (100 grams maximum) of fresh.

Cautions and contraindications: This is generally a safe herb, if you know the source, although it should be used with caution in the long term—especially in cases of a weak, cold digestive system.

Herb and drug interactions: None known.

References

C. J. Ottenhof, A. Faz Cano, J. M. Arocena, et al., "Soil Organic Matter from Pioneer Species and Its Implications to Phytostabilization of Mined Sites in the Sierra de Cartagena (Spain)," *Chemosphere* 69, no. 9 (November 2007): 1341–50.

M. S. Miao, L. Y. Gu, X. Y. Fang, et al., "Effect of *Phragmites communis* Polysaccharide on the Aged-model Mice," *Zhongguo Zhong Yao Za Zhi* 29, no. 7 (2004): 673–75.

P. Schroeder, D. Daubner, H. Maier, et al., "Phytoremediation of Organic Xenobiotics— Glutathione Dependent Detoxification in Phragmites Plants from European Treatment Sites," *Bioresource Technology* 99, no. 15 (October 2008): 7183–91.

M. Braeckevelt, G. Mirschel, A. Wiessner, et al., "Treatment of Chlorobenzene-contaminated Groundwater in a Pilot-scale Constructed Wetland," *Ecological Engineering* 33, no. 1 (May 1, 2008): 45–53.

S. Chaturvedi, R. Chandra, and V. Rai, "Multiple Antibiotic Resistance Patterns of Rhizospheric Bacteria Isolated from *Phragmites australis* Growing in Constructed Wetland for Distillery Effluent Treatment," *Journal of Environmental Biology* 29, no. 1 (January 2008): 117–24.

C. C. Carias, J. Novais, and S. Martins-Dias, "Are *Phragmites australis* Enzymes Involved in the Degradation of the Textile Azo Dye Acid Orange 7?" *Bioresource Technology* 99, no. 2 (January 2008): 243–51.

N. Marmiroli, M. Marmiroli, and E. Maestri, "Phytoremediation and Phytotechnologies: A Review for the Present and the Future," *NATO Science Series 4, Earth and Environmental Sciences* 69 (2006): 403–16.

D. Baldantoni, A. Alfani, P. Di Tommasi, et al., "Assessment of Macro and Microelement Accumulation Capability of Two Aquatic Plants," *Environmental Pollution* 130, no. 2 (2004): 149–56.

T. Vanek, A. Nepovim, R. Podlipna, et al., "Phytoremediation of Explosives in Toxic Wastes," *NATO Science Series 4, Earth and Environmental Sciences* 69 (2006): 455–65.

C. Mant, S. Costa, J. Williams, et al., "Phytoremediation of Chromium by Model Constructed Wetland," *Bioresource Technology* 97, no. 15 (October 2006): 1767–72.

P. Schroder, H. Maier, and R. Debus, "Detoxification of Herbicides in *Phragmites australis*," *Zeitschrift fur Naturforschung—A Journal of Biosciences* 60, no. 3–4 (March–April 2005): 317–24.

N. A. Ali, M. P. Bernal, and M. Ater, "Tolerance and Bioaccumulation of Cadmium by *Phragmites australis* Grown in the Presence of Elevated Concentrations of Cadmium, Copper, and Zinc," *Aquatic Botany* 80, no. 3 (November 2004): 163–76.

W. K. Chu, M. H. Wong, and J. Zhang, "Accumulation, Distribution and Transformation of DDT and PCBs by *Phragmites australis* and *Oryza sativa* L.: I. Whole Plant Study," *Environmental Geochemistry and Health* 28, no. 1–2 (February 2006): 159–68.

A. Y. Muratova, O. V. Turkovskaya, T. Hubner, et al., "Studies of the Efficacy of Alfalfa and Reed in the Phytoremediation of Hydrocarbon-polluted Soil," *Applied Biochemistry and Microbiology* 39, no. 6 (November–December 2003): 599–605.

H. Deng, Z. H. Ye, and M. H. Wong, "Accumulation of Lead, Zinc, Copper and Cadmium by 12 Wetland Plant Species Thriving in Metal-contaminated Sites in China," *Environmental Pollution* 132, no. 1 (November 2004): 29–40.

A. Massacci, F. Pietrini, and M. A. Iannelli, "Remediation of Wetlands by *Phragmites australis*—The Biological Basis," *Minerva Biotecnologica* 13, no. 2 (June 2001): 135–40.

S. T. Gregory and E. G. Nichols, "Differences in Sediment Organic Matter Composition and PAH Weathering between Non-vegetated and Recently Vegetated Fuel Oiled Sediments," *International Journal of Phytoremediation* 10, no. 6 (2008): 473–85.

N. A. Ali, M. P. Bernal, and M. Ater, "Tolerance and Bioaccumulation of Copper in *Phragmites australis* and Zea Mays," *Plant and Soil* 239, no. 1 (February 2002): 103–11.

L. Ederli, L. Reale, F. Ferranti, et al., "Responses Induced by High Concentration of Cadmium in *Phragmites australis* Roots," *Physiologia plantarum* 121, no. 1 (May 2004): 66–74.

P. Soudek, R. Tykva, and T. Vanek, "Laboratory Analyses of Cs-137 Uptake by Sunflower, Reed and Poplar," *Chemosphere* 55, no. 7 (May 2004): 1081–87.

S. D. Wallace, "On-site Remediation of Petroleum Contact Wastes Using Subsurface-Flow Wetlands," *Wetlands and Remediation* 2 (2002): 125–32.

S. Tischer and T. Hubner, "Model Trials for Phytoremediation of Hydrocarbon-contaminated Sites by the Use of Different Plant Species," *International Journal of Phytoremediation* 4, no. 3 (2002): 187–203.

C. Calheiros, A. Rangel, and P. Castro, "The Effects of Tannery Wastewater on the Development of Different Plant Species and Chromium Accumulation in *Phragmites australis*," *Archives of Environmental Contamination and Toxicology* 55, no. 3 (October 2008): 404–14.

C. Bragato, H. Brix, and M. Malagoli, "Accumulation of Nutrients and Heavy Metals in *Phragmites australis* (Cav.) Trin. ex Steudel and *Bolboschoenus maritimus* (L.) Palla in a Constructed Wetland of the Venice Lagoon Watershed," *Environmental Pollution* 144, no. 3 (December 2006): 967–75.

J. S. Weis and P. Weis, "Is the Invasion of the Common Reed, *Phragmites australis*, into Tidal Marshes of the Eastern US an Ecological Disaster?" *Mar Pollut Bull* 46, no. 7 (2003): 816–20.

Russian Olive, Autumn Olive
Elaeagnus spp.

Common names: Russian olive (*Elaeagnus angustifolia*), autumn olive (*E. umbellate*). Its botanical name refers to the "sacred olive tree," deriving from the ancient Greek *elaia* (olive tree) and *agnos* (sacred).

Family: Autumn olive family (Elaeagnaceae)

Related species: silverthorn (*E. pungens*), cherry Elaeagnus, goumi (*E. multiflorus*), *E. glabra, E. gonyanthes, E. oldhamii, E. viridus*

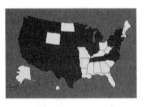

States where Russian olive is considered invasive

The Plant

Full of vitamins, minerals, antioxidants, and anticancer components, the fruits of this tree are a deeply nourishing, plentiful food source. The autumn olive has become in some cultivating circles the autumn berry, to help sell the name for wider use of this potently nutritious fruit. Strains of *Elaeagnus umbellate* are being developed for cultivation, with the brilliant rose and jewel varieties having exceptionally potent activities. While other species' fruits can be dry and mealy, they are all edible when they are fresh and are considered a valuable crop in their originating parts of the world.

> *Autumn olive might be the next Noni Gou Ji Wonderberry Cure-All Juice. Watch for it on store shelves.*

Elaeagnus is a most appropriate addition to supportive chelation therapy. Numerous vitamins, minerals, and phytonutrients can be found in the berries: vitamins C, E, A, caffeic acid, quercetin, chlorogenic acid, ferulic acid, and kaempferol, among others, have been explored individually as potent antioxidants and metal-chelating compounds. In addition, *Elaeagnus* berry is strongly active against many pathogenic influences, including cancer and multidrug-resistant strains of microorganisms.

Russian olive in the landscape sends us messages of renewal and strength to continue on in a devastated world. The tree's presence rejuvenates soil with its nitrogen-fixing capability, and it nourishes wildlife with abundant fruit. There may come a time when humans will be thankful for *Elaeagnus,* the sacred olive tree, and its widespread presence on our lands. Then the tree will be recognized, just as it was in ancient times, as a healer of disease and restorer of health, lifting the spirit with the sweet fruit it provides.

Description

Autumn and Russian olive are shrubs or small trees growing to 30 feet, with notable silvery leaves. The Russian leaves ($^1/_2$ inch wide) are thinner than the autumn olive leaves (1 inch wide) and arrange themselves alternating around the stem. In the late spring, both species bloom fragrant creamy yellow flowers. In late summer and early fall, Russian olive produces yellow-green fruits that are dry and mealy, while autumn olive has red, juicy fruits that are tart and sweet. Both species' fruit contains one seed.

Collection and Habitat

Autumn olive originates from China, Korea, and Japan, whereas Russian olive is from southeastern Europe and western Asia. Both were brought to the United States in the 1800s for many useful purposes. *Elaeagnus* was planted extensively as a supplemental food source, for ornamental use, as fast-growing windbreaks, and as a nitrogen-fixing cover crop. The dispersion of these plants for agricultural and ornamental purposes and along highways spread it far and wide throughout North America. Autumn olive is found in abundance all over the eastern half of the United States, occupying open forests, prairies, and along floodplains and roadsides. Russian olive is found more throughout the central and western United States, though it is moving eastward, and it prefers open areas, fields, streams, and roadsides.

Medicinal Uses

Parts used: Stem, leaf, and fruit

TCM: The Chinese used a variety of *Elaeagnus* species, all in very similar ways. Some include Niu nai zi (*E. umbellata*), Sha zao (*E. angustifolia*), and Yi wu gen (*E. oldhamii*)

Yi wu gen (*Radix Elaeagni*), stem and leaves; bland, cool; associated with the liver

1. Dispels wind and dampness
2. Reduces swelling and dispels stagnation

Classically in TCM, *Elaeagnus* species are used to treat cough, asthma, muscle aches, pain, weak extremities, arthritis, indigestion, diarrhea, hemorrhoids, itching, foul sores, and traumatic injury.

Ayurveda: No known uses found in ayurvedic medicine.

Western botanical: Though *Elaeagnus* species are new to the Western herbal world, agriculturalists have produced cultivars for the berries, which are being studied extensively for powerful healing compounds. See scientific studies below.

Plant Chemistry

Lycopene, beta-carotene, B-sitosterol, boron, calcium, tocopherol, ascorbic acid, alkaloids, caffeic acid, calligonine, catechin, chlorogenic acid, dihydroharman, elaeagnine, elaeagnoside, ellagic acid, epicatechin, ferulic acid, harman, kaempferol, lycopene, N-methyl-tetrahydroharmol, neo-chlorogenic acid, p-coumaric acid, quebrachitol, quercetin, sinapic acid

Elaeagnus pungens: 7-dimethylkaempferol, kumatakenin, lupeol, oleanolic acid, ursolic acid, vanillic acid, phytoene, a- and ß-cryptoxanthin

Pharmacological Actions

Antimicrobial, anticancer, antioxidant, antiplatelet aggregation, anti-inflammatory

Scientific Studies

Antimicrobial: One study reports the in vitro antibacterial activity of *E. umbellate,* with extracts (especially from the flowers) displaying broad-spectrum activity against gram-positive and -negative bacteria, methicillin-resistant *Staphylococcus aureus, Bacillus subtilis, E. coli,* and multidrug-resistant *Pseudomonas aeruginosa.*

Antioxidant: The berries of autumn olive have been found to contain high levels of antioxidants, including vitamins A, C, and E, quercetin, lycopene, and other flavonoids and essential fatty acids. The potent antioxidant lycopene is found in abundance: the berries have up to seventeen times more lycopene than the typical raw tomato, where most people get this phytochemical.

Anticancer: With their free radical–scavenging capacities, in a USDA study, autumn olive berries were found to inhibit the proliferation of human leukemia cancer cells, human lung epithelial cancer cells, and induced apoptosis of HL-60 cells in in vitro studies. A methanol extract of *E. glabra* was found in vitro to have inhibitory effects on tumor cell invasion.

Analgesic: The fruit of *E. angustifolia* was found to help mice tolerate pain. The analgesic effects were found to be based in the central nervous system, and the mechanism of pain relief was not attributed to anti-inflammatory action or through opioid pathways. The fruit was found to also induce muscle relaxation in mice.

Other Uses

The fruit can be made into jams, jellies, and preserves, dried into fruit leather, or cooked with other foods, and in Asia an alcoholic beverage is made from them. The highly nutritious fruits contain vitamins B_1, B_2, C, and E, beta-carotene, lycopene, boron, calcium, carbohydrates, protein, chromium, copper, fat, fiber, folic acid, iron, magnesium, manganese, niacin, phosphorus, potassium, tryptophan, and zinc.

Leaves of *Elaeagnus* can be used as livestock fodder. It also serves as a fast-growing windbreak and cover.

Ecological Importance

Through a symbiotic relationship with bacteria, *Elaeagnus* is able to bring nitrogen to the soil from the air and is considered a nitrogen-fixing species. *Elaeagnus* also reclaims strip-mined land and has the ability to accumulate lead and zinc from toxic soils. The plant replenishes burdened agricultural lands and protects disturbed ecosystems throughout North America.

Russian and autumn olive serve as important wildlife habitat and food providers. The fruits of *Elaeagnus* remain on the tree into winter, and the flowers also provide nectar to insects and bees.

Harvesting and Preparing

Harvest and prepare as usual. Fruits of autumn berry can be freshly juiced.

Dosage

Medium- to high-dose botanical, with higher doses for the fruit.

Decoction: The TCM dosage is 3 to 9 grams of the stem and leaves.
Cautions and contraindications: None known.
Herb and drug interactions: None known.

References

S. Y. Wang, L. Bowman, and M. Ding, "Variations in Free Radical Scavenging Capacity and Antiproliferative Activity Among Different Genotypes of Autumn Olive (*Elaeagnus umbellata*)," *Planta Medica* 73, no. 5 (May 2007): 468–77.

A. Khamzina, J. P. Lamers, and P. L. Vlek, "Nitrogen Fixation by *Elaeagnus angustifolia* in the Reclamation of Degraded Croplands of Central Asia," *Tree Physiology* (March 20, 2009).

J. Strax, "Autumn Olive, a Berry High in Lycopene," *PSA Rising—Prostate Cancer Activist News,* www.psa-rising.com/eatingwell/wild-foods/autumnolive.htm (accessed April 6, 2009).

A. Chehregani and B. E. Malayeri, "Removal of Heavy Metals by Native Accumulator Plants," *International Journal of Agriculture and Biology* 9, no. 3 (2007): 462–65.

M. S. Sabir, D. S. Ahmad, I. M. Hussain, and K. M. Tahir, "Antibacterial Activity of *Elaeagnus umbellata* (Thunb.) a Medicinal Plant from Pakistan," *Saudi Med J* 28, no. 2 (2007): 259–63.

S. Y. Wang, L. Bowman, and M. Ding, "Variations in Free Radical Scavenging Capacity and Antiproliferative Activity among Different Genotypes of Autumn Olive (*Elaeagnus umbellata*)," *Planta Med* 73, no. 5 (2007): 468–77.

M. Ramezani, H. Hosseinzadeh, and N. Daneshmand, "Antinociceptive Effect of *Elaeagnus angustifolia* Fruit Seeds in Mice," *Fitoterapia* 72, no. 3 (2001): 255–62.

A. Ahmadiani, J. Hosseiny, S. Semnanian, M. Javan, F. Saeedi, M. Kamalinejad, and S. Saremi, "Antinociceptive and Anti-inflammatory Effects of *Elaeagnus angustifolia* Fruit Extract," *Journal of Ethnopharmacology* 72, no. 1–2 (2000): 287–92.

L. H. Li, I. K. Baek, J. H. Kim, et al., "Methanol Extract of *Elaeagnus glabra,* a Korean Medicinal Plant, Inhibits HT1080 Tumor Cell Invasion," *Oncol Rep* 2, no. 2 (2009): 559–63.

H. Hosseinzadeh, M. Ramezani, and N. Namjo, "Muscle Relaxant Activity of *Elaeagnus angustifolia* L. Fruit Seeds in Mice," *J Ethnopharmacol* 84, no. 2–3 (2003): 275–78.

Scotch Broom
Cytisus scoparius

Common names: Scotch broom, Scots broom

Family: Pea family (Fabaceae)

Related species: There are ten species of *Cytisus* including White Spanish broom (*C. multiflorus*) and striated broom (*C. striatus*)

States where Scotch broom is considered invasive

The Plant

Held in the highest esteem by many lords of the Old World, Scotch broom became a badge of honor worn by kings and noblemen. According to Maud Grieve, "the Scottish clan of Forbes wore [Broom] in their bonnets when they wished to arouse the heroism of their chieftains"[32] and as a badge of courage when going to battle. Geoffrey of Anjou "thrust it into his helmet," and declared, "This golden plant, rooted firmly amid rock, yet upholding what is ready to fall, shall be my cognizance. I will maintain it on the field, in the tourney (tournament of jousting), and in the court of justice."[33] Broom became an official ally of heroic individuals: It was embedded in seals and tombs of kings and has a long tradition of standing by humans.

The heroic nature of Scotch broom is highlighted in the potent medicine it provides and in its role as an ecological restorer of devastated terrain. Scotch broom is full of antioxidants, acts as a nervous system relaxant, and has liver-cleansing capabilities said to be comparable to those of milk thistle. This plant protects disrupted areas from erosion, cleans toxic defilements from the soil, and replenishes the land with nurturing nitrogen-fixing effects.

> *Scotch broom simply follows on the heels of environmental disturbance, covering soil that has been laid bare by logging, road building, development, etc. The plant is not only an earth healing nitrogen fixer but also a potent medicinal plant used to strengthen and regulate the heartbeat, treating arrhythmia and edema associated with congestive heart failure, [and the] source of the medicinal compound Sparteine.*
>
> RICHO CECH, HORIZON HERBS

Description

Broom is a perennial shrub (3–6 ft. tall) with deep roots to reach water sources in dry and well-drained soils. The plant creates a mesmerizing cloud of yellow along roadsides in the summertime, with thousands of individual bright yellow flowers that look similar to sweet pea flowers. In the late summer and early autumn, the flowers form fuzzy, flat seedpods 1 to 2 inches long that explode when ripe, spreading the seeds far and wide. The seedpods can remain viable for up to thirty years in the soil.

Collection and Habitat

Scotch broom is native to the Mediteranean region and was brought to this continent in the early 1800s for beautiful ornamental purposes and for its vigorous growth in difficult terrain. It helps bind sandy soils and is a good sheltering plant for seaside growth. It was further planted throughout the land for erosion control and soil improvement. Today, Scotch broom is found abundantly on roadsides, sand dunes, and timbered areas in North America, inhabiting much of the West Coast and into Alaska and throughout much of the eastern United States.

Medicinal Uses

Parts used: Flowering tops, and seeds
TCM: No known reference in TCM.
Ayurveda: No known reference in ayurveda.
Western botanical: *King's American Dispensatory* describes Scotch broom as diuretic in small doses and emetic and cathartic in large doses. It has been used for edema, cardiac arrhythmia, nervous cardiac complaints,

low blood pressure, heavy menstruation, hemorrhaging after birth and to stimulate contractions, for bleeding gums, hemophilia, gout, rheumatism, sciatica, gall and kidney stones, enlarged spleen, jaundice, bronchial conditions, and snakebites.

The herb has been approved by the German Commission E for circulatory disorders and hypertension.

Plant Chemistry

Beta-sitosterol, chrysanthemexanthin, cytisine, dopamine, epinine, esculetin, flavoxanthin, furfurol, genistein, genisteine, gibberellin-A, hydoxylupanine, hydoxytyramine, hyperoside, isoquercitrin, isosparteine, linoleic acid, lupanine, luteolin, methyloxytyramine, orientin, oxytyramine, quercetin, sarothamnine, scoparin, sparteine, spirasoside, tannin, taraxanthin, tyramine, vitexin, xanthophyll, 12-dehydrosparteine, 17-oxosparteine

Pharmacological Actions

Antioxidant, antidiabetic, hypnotic, sedative, cardiovascular relaxant

Scientific Studies

Antioxidant: In one study, Scotch broom was found to protect the liver against oxidative stress in rats, with results comparable to the compound silymarin, the main liver-supporting component of milk thistle. The antioxidant effects of *Cytisus scoparius* were also shown to contribute potentially to a rat's ability to handle stress and manage anxiety, which ultimately increases oxidative stress on the body. Due to flavonoids and other polyphenolic compounds, the "alcoholic extract of aerial part of *Cytisus scoparius* is a potential source of natural antioxidant,"[34] and exhibits scavenging abilities of free radicals.

Other Uses

The young blossoms can be pickled in vinegar and are likened to capers. The tender green tops were used as a replacement for hops in beer, for both the bitterness and the intoxication. The seed was also roasted as a coffee substitute for early settlers.

Cytisus pullilans is said to have excellent forage value, especially for livestock and poultry. When the stem grows to a sufficient size, it is beautifully veined and hard and is a valuable material for carpentry and veneering. As its name implies, Scotch broom provides fiber for brooms, and—from twigs and branches—fiber for baskets. The bark and shoots can be used to make fiber for paper and cloth. In addition, the tannins in the plant have made it useful in tanning leather.

Ecological Importance

Along with a few other invasives, *Cytisus striatus* was found to accumulate the pesticide hexachlorocyclohexane (HCH) in contaminated soils of northwestern Spain, and it was also found to remediate the element manganese (Mn).

As it thrives on disturbed land and roadside waste areas, providing valuable erosion control, Scotch broom also improves the soil with its nitrogen-fixing capabilities. There are some discrepancies as to the actual invasiveness of this plant: a USDA Forest Service study suggests "that intact forest communities are relatively non-susceptible to invasion,"[35] which, as pointed out by Richo Cech, "means that the real invader is the human, not the plant."[36]

In Australia a study points out that the widespread occurance of Scotch broom there was attributed to the introduced honeybee, with the plant unable to spread until the honeybee entered the environment. This is one more piece of the invasive plant puzzle that shapes the plant's presence in the landscape.

Harvesting and Preparing

Harvest and prepare the flowering tops of broom as described.

Dosage

Low- to medium-dose botanical.

Decoction: For a strong decoction, cook 1 ounce of the tops in 1 pint of water for 10 minutes. Dosage for the seeds consists of 10 to 15 grains, pulverized and taken with fluids.

Tincture: 15 to 30 drops, though some consider a tincture inferior to the decoction.

Cautions and contraindications: Avoid use during pregnancy. Some adverse effects include a staggering gait, impaired vision, vomiting, and sweating. Some individuals (up to 10 percent) have an impaired ability to eliminate sparteine, leading to toxicities.

Herb and drug interactions: When taking, avoid MAO inhibiters and drugs that inhibit the metabolism of the component sparteine, which can lead to toxicity. In vitro studies have shown the relevant inhibitors to be quinidine, haloperidol, amiodarone, chlorpheniramine, cimetidine, clomipramine, fluoxetine, methadone, paroxetine, and ritonavir.

References

S. Raja, K. F. Ahamed, V. Kumar, et al., "Antioxidant Effect of *Cytisus scoparius* Against Carbon Tetrachloride Treated Liver Injury in Rats," *Journal of Ethnopharmacology* 109, no. 1 (2007): 41–47.

J. Nirmal, C. S. Babu, T. Harisudhan, et al., "Evaluation of Behavioural and Antioxidant Activity of *Cytisus scoparius* Link in Rats Exposed to Chronic Unpredictable Mild Stress," *BMC Complementary Alternative Medicine* 8 (2008): 15.

R. Calvelo Pereira, M. Camps-Arbestain, B. Rodríguez Garrido, et al., "Behaviour of Alpha-, Beta-, Gamma-, and Delta-hexachlorocyclohexane in the Soil-plant System of a Contaminated Site," *Environ Pollut* 144, no. 1 (November 2006): 210–17.

T. B. Harrington, "Factors Influencing Regeneration of Scotch Broom (*Cytisus scoparius*)," Pacific Northwest Research Station and USDA Forest Service, www.ruraltech.org/video/2006/invasive_plants/pdfs/NHS_Hall/17_harrington.pdf (accessed May 6, 2009).

V. De la Fuente, L. Rufo, N. Rodríguez, R. Amils, and J. Zuluaga, "Metal Accumulation Screening of the Río Tinto Flora (Huelva, Spain)," *Biol Trace Elem Res* (August 8, 2009), doi: 10.1007/s12011-009-8471-1.

S. R. Simpson, C. L. Gross, and L. X. Silberbauer, "Broom and Honeybees in Australia: An Alien Liaison," *Plant Biol* (Stuttg) 7, no. 5 (2005): 541–44.

R. Sundararajan, N. A. Haja, K. Venkatesan, et al., "*Cytisus scoparius* Link—a Natural Antioxidant," *BMC Complement Altern Med,* no. 6 (March 16, 2006): 8.

Siberian Elm
Ulmus pumila

Common names: Siberian elm, dwarf elm, littleleaf elm

Family: Elm family (Ulmaceae)

Related species: Slippery elm, red elm, Indian elm (*U. rubra* or *U. fulva*), American elm (*U. americana*), Chinese elm (*U. parviflora*)

States where Siberian elm is considered invasive

The Plant

Many older people will remember the grand elm trees lining neighborhood streets when rock 'n' roll was born. Then thoughts move on to the disastrous fungal plague—Dutch elm disease—that took down nearly every single one of these giant trees, with the help of a simple beetle.

> *How could such a lowly creature do such monstrous harm? The tale of the giant being taken down by the diminutive is constantly repeated in our teachings: David versus Goliath; Knight versus Dragon; Fungus versus Tree; Bacteria versus Human; Human versus Earth.*

Siberian elm is often attacked for spreading throughout the land. This, however, presents an interesting contradiction: scientists rely heavily on the germ plasm of Siberian elm breeding with endangered American elm to produce cultivars that are resistant to Dutch elm disease.

A relevant story, found in the journal *Nature,* cites a major contribution to the virulence of Dutch elm disease that spread throughout

Europe just after it passed through America and was carried across the sea. It was found that the English elm tree (*U. procera*), so widespread throughout the continent, was derived from a single clone of a two-thousand-year-old tree that was planted on the Iberian peninsula by the Romans. The elm tree chosen for propagation did not set seed and was efficient at vegetative reproduction. The massive, large-scale distribution of the tree led to its downfall as well, creating a less adaptable gene pool in the elm's diversity and wiping out twenty-five million trees in 1970s Britain alone. "The preponderance of this susceptible variety may have favoured a rapid spread of the disease."[37] Fortunately though, the Siberian elm is resistant to Dutch elm disease and could prove itself to be the savior of native elm species through genetics and medicinal attributes.

The ever-expanding nature of Siberian elm (*Ulmus pumila*) throughout the United States might also help save another native elm species. It makes an appropriate substitute for the powerful, long-sought and now endangered medicinal tree known as the slippery elm (*U. rubra* or *U. fulva*). This tree's medicine has been sought for ages for its cooling and soothing effects on mucous membranes and inflamed tissues of the body. Over the course of the past century, slippery elm has been overharvested for medicine, weakened by disease, and burdened with the pressures of urbanization, logging, and mining. One way we can help this species from demise is by making use of the rampantly spreading Siberian elm trees as medicine.

Description

To the untrained eye, Siberian elm is easily confused with North American varieties of *Ulmus*. The distinction of Siberian elm is its smaller leaves ($1/2$–1 inch wide by 2 inches long), which alternate along the stem and are toothed around the edges, ribbed and shiny, and the typical elm shape, being oval and tapered to a point. Many stems can emerge from the trunk, which gives the tree an overall round appearance. In open spaces, it can grow up to sixty feet tall. The bark of mature trees is light gray-brown and furrowed. Siberian elm's greenish red to brown petalless flowers bloom in the summer and then give way to clusters of brown, round-winged seed cases in the fall, each with a single seed.

Collection and Habitat

Siberian elm was brought to America in the 1860s and planted as a wind-break throughout the Midwest. Originally from Siberia, northern China, and Korea, *Ulmus pumila* tolerates extreme conditions of intense wind, bitter cold, and poor soil. Siberian elm trees and shrubs are now ubiquitous throughout North America as a dominant succession species that quickly moves into old fields, pastures, and disturbed areas, as well as growing along roadways and streams.

Medicinal Uses

TCM: Wu yi (*Ulmus macroparpa*), fruit or seed; acrid, bitter, warm; associated with the spleen and stomach

Kills parasites and reduces digestive stagnation. Classically, Siberian elm is used for roundworm, tapeworm, tinea infection, and other parasitic infestations. Chinese elm, Lang yu pi (*U. parvifolia*), is used for complaints including skin abscesses, swellings, wounds, infections, bladder ailments, diarrhea, gonorrhea, insomnia, ulcers, and fever.

Ayurveda: No uses found.

Western botanical: Elm species across continents have been used for medicine. The most notable is slippery elm (*U. rubra*): The inner bark is used for soothing irritated, inflamed mucous membranes of the stomach, intestines, and lungs. Inner bark powder can be used externally in poultices or internally for a mucilaginous drink. Slippery elm is also one ingredient found in the original Ojibwa formula for cancer, now known as the Essiac herbal formula for cancer treatment.

Plant Chemistry

Arjunolic acid, camaldulenic acid, dehydroulmudiol, epifriedelanol, friedelin, mansonone, maslinic acid, oleanolic acid, ulmudiol, ulmuestone

Pharmacological Actions

Antiparasitic, antifungal, antidotal, diuretic, antilithic (removes stones), expectorant, demulcent, soporific

Scientific Studies

Antimicrobial: An extract of Wu yi (*U. macroparpa*) was found to kill roundworms in pigs and inhibit the growth of many pathogenic fungi.

Anticancer: Two compounds were isolated in order to find their effects on human tumor cells in vitro. Mansonone E and F were found to have antiproliferation activities on four cell lines: human cervical cancer, human malignant melanoma, human breast cancer, and human histiocytic lymphoma.

Antihypertensive: The root bark of *U. macrocarpa* extract was studied to evaluate its cardiovascular effects and antioxidant activity. Using rats, the extract showed significant vascular relaxation, decreased blood pressure with long-term treatment, and provided resistance to hydrogen peroxide–mediated oxidative stress on myocardial cells.

Antiulcer: The root bark of elm was found to help heal ulcers and regenerate tissue growth in a study on ulcerated rats.

Other Uses

North American Indian tribes ate the dried, powdered inner bark of slippery elm raw or boiled with fat for the flavor and used it as a preservative. It is easily digestible and nutritious—perfect for infants and the seriously ill mixed with milk. Asian and European elms have been eaten in similar ways, including the use of the young leaves and fruits. The leaves and young shoots are suitable food for livestock.

> *When crops failed or long severe winters exhausted food supplies, Indians and pioneers alike were often saved from starvation by the use of Sweet Elm bark. This emergency source of food had the advantage of being available when all other sources of food had failed. The use of Sweet Elm bark as food spread with the early colonies until the day when the vast forests were converted into farm lands.*
>
> THE HERBALIST ALMANAC, FROM EAT THE WEEDS
> BY BEN CHARLES HARRIS

Elm is used to make furniture, baskets, boxes, and crates. The strong inner bark is made into mats and ropes. The wood is used for biomass energy.

Ecological Importance

One study set out to determine the viability of naturally attuned plants growing around industrial areas of Beijing, China. Concentrations of the heavy metals, iron, manganese, aluminum, zinc, lead, nickel, chromium, and arsenic were found in the soil surrounding a steel factory, and Siberian elm (*U. pumila*) was found to be a strong to moderate accumulator of the toxins and was considered useful for remediating purposes.

In studying accumulation rates of perchlorate by numerous aquatic and terrestrial plants, scientists found elm (*U. parvifolia*) to be one of the most efficient species at accumulating this pollutant. In Iran, *U. carpinifolia* was evaluated and subsequently recommended for its ability to remove lead, cadmium, and copper from metal-polluted waters.

Early spring flowers of elm trees provide abundant pollen and nectar for bees and insects.

Harvesting and Preparing

Harvest and prepare the various parts of Siberian elm as described. The inner bark is best harvested in the spring and can be prepared externally for poultice.

Dosage

Medium-dose botanical. Dosage for seeds and fruit are lower than for the inner bark.

Decoction: Two ounces of bark in one quart of water, steeped for tea.
TCM dosage: 3 to 10 grams of seeds taken as pills or powder.
Cautions and contraindications: None known.
Herb and drug interactions: None known.

References

Dong Wang, Mingyu Xia, and Zheng Cui, "New Triterpenoids Isolated from the Root Bark of *Ulmus pumila* L.," *Chemical & Pharmaceutical Bulletin* 54 (2006): 775–78.

D. Wang, M. Xia, Z. Cui, et al., "Cytotoxic Effects of Mansonone E and F Isolated from *Ulmus pumila*," *Biological and Pharmaceutical Bulletin* 27, no. 7 (2004): 1025–30.

J. E. Zalapa, J. Brunet, and R. P. Guries, "Genetic Diversity and Relationships among Dutch Elm Disease Tolerant *Ulmus pumila* L. Accessions from China," *Genome* 51, no. 7 (July 2008): 492–500.

K. S. Oh, S. Y. Ryu, B. K. Oh, et al., "Antihypertensive, Vasorelaxant, and Antioxidant Effect of Root Bark of *Ulmus macrocarpa*," *Biological and Pharmaceutical Bulletin* 31, no. 11 (November 2008): 2090–96.

Y. K. Na and H. S. Hong, "The Effects of the Ulmus Root-bark Dressing in Tissue Regeneration of Induced Pressure Ulcers in Rats," *Taehan Kanho Hakhoe Chi* 36, no. 3 (June 2006): 523–31.

L. Gil, P. Fuentes-Utrilla, A. Soto, et al., "Phylogeography: English Elm Is a 2,000-year-old Roman Clone," *Nature* 431, no. 7012 (October 28, 2004): 1053.

K. Tan, T. A. Anderson, M. W. Jones, et al., "Accumulation of Perchlorate in Aquatic and Terrestrial Plants at a Field Scale," *Journal of Environmental Quality* 33, no. 5 (September–October 2004): 1638–46.

Y. J. Liu, H. Ding, and Y. G. Zhu, "Metal Bioaccumulation in Plant Leaves from an Industrious Area and the Botanical Garden in Beijing," *Journal of Environmental Sciences* (China) 17, no. 2 (2005): 294–300.

M. R. Sangi, A. Shahmoradi, J. Zolgharnein, et al., "Removal and Recovery of Heavy Metals from Aqueous Solution Using *Ulmus carpinifolia* and *Fraxinus excelsior* Tree Leaves," *Journal of Hazardous Materials* 155, no. 3 (July 15, 2008): 513–22.

Tamarisk
Tamarix spp.

Common names: tamarisk, salt cedar

Family: Tamarisk family (Tamaricaceae)

Related species: There are at least nine species of tamarisk growing in North America, including *Tamarix ramosissima, T. africana, T. aphylla, T. chinensis, T. gallica,* and *T. parviflora*

States where tamarisk is considered invasive

The Plant

Tamarisk has been demonized by the claim that it uses more than twice as much water annually as all the cities in southern California, placing it in direct competition with humans for the most limiting resource in the southwestern United States. This plant is capable of reaching deep-water reserves along rivers and in flood plains, yet the expansion of tamarisk only surged in the early 1900s, coincidentally at the same time massive damming projects were underway in the desert to divert vast amounts of water to serve the needs of golf courses, green lawns, and farms in the Southwest. The cost of mass eradication efforts of tamarisk are beginning to show that in removing it, there is a severe impact on erosion and increased sedimentation. We are realizing that the removal of tamarisk may permanently reduce reservoir capacity and increase evaporation rates, thereby further depleting water supplies. The success of this plant is due to its ability to reach low levels of groundwater, when the native flora requires a higher water table. As the water level increases, tamarisk's aboveground biomass and the salt deposits this tree exudes are decreased, thereby indicating human's excess water use in an area. A U.S. Geological Survey study found that efforts to eradicate this plant do not lead to increased water supplies and are not a successful course of action for revegetation efforts. The agriculture department of Texas A&M concurs, with "water salvage estimates show[ing] a significant reduction in system water loss after Salt Cedar treatment."[38] In addition, eradication does not reduce the salinity of the soils where tamarisk grows, and not all wildlife prefer native plant habitat over the tamarisk ecosystem. This clearly demonstrates that this tree is not harming the environment and was, in fact, an ecological intervention to remedy our human woes and preserve our precious resource.

Salt cedar (tamarisk) is a long-used remedy across the globe. Many species in the Tamarisk family have been used in Europe, Africa, Asia, and now America, and all provide medicine that has effectively dealt with infections, dysentery, and human toxic accumulations. Classically, tamarisk is used for various pathogenic influences such as influenza, measles, chicken pox, syphilis, and dysentery; and for painful inflammations such as arthritis, piles, fissures, hemorrhages, sore throat, stomatitis, and toothache. The nature of tamarisk—excreting salt and toxic defilements through its leaves—reveals the significance of its medicine in relieving human ecology of toxic pathogenic influences, especially through the skin. These traditional uses have now been verified through scientific measures that detected the plant's antimicrobial, antioxidant, and liver detoxifying abilities (through the compounds quercetin, tamarixetin, tannins, and sodium sulfate, among others). Tamarisk is a remediator of the body and healer of illness.

Description

Tamarisk is tall shrub or small tree (10–15 feet tall) that can form dense colonies with entangling stems that can grow to 30 feet across. The trees look like cedar, with fine, scaly leaves, which from a distance create tamarisk's distinctive feathery appearance. The bark is smooth and reddish brown to purplish dark brown in color. From early spring to late fall it can be found producing plumes of tiny white-to-pink flowers with four or five petals that create dense clusters (2–3 inches long) emerging from the tips of branches. A tiny seed forms within a capsule that has a tuft of hair to catch the wind. *Tamarix aphylla* grows larger (up to 50 feet) and is evergreen, while the other species are deciduous.

Collection and Habitat

Originally from the Mediterranean region eastward across the Middle East to China and Japan, tamarisk was imported to North America in the early 1800s for erosion control along watercourses, for wind breaks, and for landscaping. It tolerates very extreme conditions and difficult soils, surviving fires, water submersion for over two months, drought, and saline soil conditions. Spread of tamarisk began to surge in the early 1900s after the damming of many rivers and is now a dominant species along riparian areas of the Southwest and has moved north and east, occupying most of the United States and southern Canada.

Medicinal Uses

Parts used: Branches with leaves, bark, galls

TCM: Xi he liu (*T. chinensis*), "western river willow," branches, leaves; salty, sweet, warm; associated with lungs, stomach, heart

1. Clears heat, resolves toxins
2. Drains dampness
3. Induces sweating and vents rashes

Classically, tamarisk is used to treat influenza, measles, chicken pox, rheumatoid arthritis, and alcoholic intoxication.

Ayurveda: Jhavuka, Shaavaka (*T. orientalis*), root, leaf, galls

Classically, tamarisk is used to treat piles, fissures, diarrhea, dysentery, hemorrhages, sore throat, stomatitis, and toothache.

Western botanical: Used in southern Europe as a compress to stop bleeding wounds. James Duke cites biblical references for tamarisk, using it for eczema, infertility, impotence, ophthalmia, psoriasis, splenitis, and syphilis. In other sources, it also has been found useful in leucoderma, spleen troubles, and eye diseases.

Plant Chemistry

Tamarixin, isoquercitrin, rhamnocitrin, isorhamnetin, isoferulic acid, hexadecanoic acid, docosane, germacrene D, fenchyl acetate, Benzyl benzoate, phytyl acetate, sodium sulfate, tannins, gallic acid, ellagic acid, dehydrodigallic acid, juglanin, quercetin, tamarixetin, tamarixinol, B-sitosterol

Pharmacological Actions

Antimicrobial, astringent, antidiarrhetic, stimulates appetite and perspiration, diuretic

Scientific Studies

Antioxidant: Tamarisk contains the antioxidant compounds quercetin, tamarixetin, and other polyphenolic substances, which have been shown to scavenge for free radicals and assist in heavy-metal chelation therapy.

Antimicrobial: An in vitro study of *Tamarix boveana* and its volatile oils revealed antibacterial activity against five strains, including *Staphylococcuss aureus, S. epidermidis, E. coli, Micrococcus luteus,* and hospital *Salmonella typhimurium.*

Liver cirrhosis: In a randomized, double-blind, placebo-controlled study, an herbal formula containing *T. gallica* was found to possess hepatoprotective actions in cirrhotic patients after six months of treatment, which was attributed to the diuretic, anti-inflammatory, antioxidant, and immunomodulating properties of the herbs.

Other Uses

Its significant tannins make it useful in the tanning industry. It is also used for dyeing fabrics; it imparts a pink and purple color. Its wood is used for construction and fuel. It is planted as a windbreak in saline and hard-to-grow environments (as along the seacoast). In the Middle East, the stems of *Tamarix mannifera* exude a sweet substance (manna), which results from the insect stings in the stem. (Probably, antioxidant flavanoids are released to heal the wound.)

Ecological Importance

Cleans the pollutants lead, cadmium, copper, arsenic, sodium, and perchlorate (an oxidant with connections to thyroid disorders) and in general is considered a very qualified species for wastewater purification. Tamarisk is also a unique species in that it uses a bioremedition process termed *phytoexcretion,* by which the plant releases contaminants through its foliage. Tamarisk also forms dense stands, which protects areas by preventing erosion along waterways and by acting as a windbreak. Tamarisk can colonize heavily saline soils where little else grows.

Tamarisk supplies plentiful nectar for honeybees and insects as well as providing shelter for doves and the endangered southwestern willow flycatcher, which prefers salt cedars for nesting.

Harvesting and Preparing

Harvest and prepare as described for both internal and external use. The twigs and leaves can be placed in a bath to treat measles and other skin ailments.

Dosage

Medium- to low-dose botanical.

Decoction: The TCM dosage is 3 to 9 grams.

Cautions and contraindications: Contraindicated for internal use when there are fully erupted measles and when there is profuse sweating and/or severe weakness. Overdose can cause vomiting, dizziness, excess sweating, spasms, lowered blood pressure, and, in extreme cases, shock and loss of consciousness.

References

C. R. Hart, L. D. White, A. McDonald, et al., "Salt Cedar Control and Water Salvage on the Pecos River, Texas, 1999–2003," *Journal of Environmental Management* 75, no. 4 (2005): 399–409.

E. T. Urbansky, M. L. Magnuson, C. A. Kelty, et al., "Perchlorate Uptake by Salt Cedar (*Tamarix ramosissima*) in the Las Vegas Wash Riparian Ecosystem," *Sci Total Environ* 256, no. 2–3 (2000): 227–32.

E. Manousaki, J. Kadukova, N. Papadantonakis, et al., "Phytoextraction and Phytoexcretion of Cd by the Leaves of *Tamarix smyrnensis* Growing on Contaminated Non-saline and Saline Soils," *Environ Res* 106, no. 3 (2008): 326–32.

H. F. Huseini, S. M. Alavian, R. Heshmat, et al., "The Efficacy of Liv-52 on Liver Cirrhotic Patients: A Randomized, Double-blind, Placebo-controlled First Approach," *Phytomedicine* 12, no. 9 (2005): 619–24.

J. Kadukova, E. Manousaki, and N. Kalogerakis, "Pb and Cd Accumulation and Phyto-excretion by Salt Cedar (*Tamarix smyrnensis* Bunge)," *Int J Phytoremediation* 10, no. 1 (2008): 31–46.

N. Sultanova, T. Makhmoor, Z. A. Abilov, et al., "Antioxidant and Antimicrobial Activities of *Tamarix ramosissima*," *Journal of Ethnopharmacology* 78, no. 2–3 (2001): 201–5.

D. Saidana, M. A. Mahjoub, O. Boussaada, et al. "Chemical Composition and Antimicrobial Activity of Volatile Compounds of *Tamarix boveana* (Tamaricaceae)," *Microbiological Research* 163 (2008): 445–55.

M. Adrover, A. L. Forss, G. Ramon, et al., "Selection of Woody Species for Wastewater Enhancement and Restoration of Riparian Woodlands," *J Environ Biol* 29, no. 3 (2008): 357–61.

Thistle
Cirsium spp.

Common names: Canada thistle, creeping thistle, way thistle, California thistle, cirsium (*C. arvense*); bull thistle, common thistle, spear thistle (*C. vulgare*)

Family: Aster family (Asteraceae)

Related species: There are ninety-five species of *Cirsium* that grow throughout the world including field thistle (*C. discolor*), Japanese thistle (*C. japonicum*), meadow thistle (*C. scariosum*), *C. setidens,* and technically not the same species but related, blessed thistle (*Cnicus benedictus*)

States where Canada thistle is considered invasive

The Plant

Thistle is widespread throughout the world, and they are a nuisance to those who wish to use the land where they grow. Those people who are in sight of the plant know the prevailing nature of thistle, with its sharp thorns that seem to stab painfully from a distance. The energetics of thistle palpably spring from the plant and tell much about its potency for both human healing and ecological revitalization.

Various species of thistle have been used in numerous herbal traditions in similar ways for thousands of years. The most famous of these is the use of milk thistle as a potent medicinal for liver health and detoxification (though this plant is a different species from the one discussed here). The most persistently expanding thistles of the *Cirsium* clan in North America are Canada thistle (*Cirsium arvense*) (not from Canada) and bull thistle (*Cirsium vulgare*). (Star thistles are also a different species: *Centaurea.*) In fact, the wildly expansive thistles and their seeds make a good substitute for milk thistle. In general, they have a universal effect on the liver, and they help aid in the detoxification processes of the body.

Anger is known to be an intimate expression of the liver, and being around this plant often provokes a livid reaction that is often coupled with profanities even in the most civil and refined people. This response gets the blood boiling, circulating it to the face and eyes with redness and heat as the manifestation of the liver flares to the point of explosion. The movement of the blood speeds up and spreads out to the extremities, and the fists are clenched with rage. Because the circulation system is essentially a closed loop, all blood flows with greater rates during a burst of anger and is sent through the liver for filtering—thereby the action of the liver is instigated. The liver, as we have seen, is known as the purifier of the blood. It works hard to remove metabolism by-products, toxins, and poisonous influences from the body.

No wonder getting angry can be so tiring!

The environments that thistle rampantly inhabits are generally in need of rest and rejuvenation. Overused and depleted agricultural and rangelands are the areas most in need of this plant—and these areas are where it grows. It keeps foraging cattle from such lands and discourages farmers with its virulence. Thistle's roots aerate the generally hard soil of improperly managed rangeland, and over time, the plant increases biomass to restore and conserve the topsoil from blowing away. Thistle is a nutritious, power food for the land and the body, restoring and expanding the desolate spirit even in the harshest of circumstances.

Rosemary Gladstar recounts one of her tales of wild living:

> I lived on all varieties of thistles one summer when I was horseback riding from Sonoma County to northern California (the Trinity Alps). It was a drought and there was very little food available as we were wild crafting our own food as we rode. I ate thistles in every form you can imagine and they are actually quite tasty; the original artichoke. Thistle stems can be boiled and eaten, or eaten raw, and the leaves, with the barbs taken off, can be steamed or eaten raw, while the roots are best steamed. The heart of the flower also provides little taste treats. All the thistles are good tasting and are very good for you, too.[39]

Description

All thistles are known for their spiny leaves and sharp presence. The perennial Canada thistle grows as dense colonies of individual plants (1.5–5 feet tall), with the distinct ability of this *Cirsium* species to spread by vigorous root runners that produce clones around the mother plant. The leaves vary in size and alternate around a hairy stem. In general they are irregularly lobed with numerous spines along the edges. The male and female flowers occur on separate plants and bloom throughout the summer, each flower being small (½ to ¾ inch across) and violet to pink in color. The seed heads turn brown and release scores of tiny seeds that disperse by wind.

Bull thistle is a biennial plant that creates a rosette of spiny leaves the first year, which then turns into a grand presence of spiky leaves and stems with numerous rose-pink flowers. The plant can grow to 8 feet tall with stems branching out to 3 to 4 feet wide. The flowers (1 inch) bloom throughout the summer and into fall, and soon thereafter many tiny and hairy wind-blown seeds form in the seed head.

Collection and Habitat

Many thistle species were brought over with early settlers from Europe and Asia in the early 1600s, most likely unintentionally through livestock feed and agricultural seed. Thistles are now found throughout North America, growing in a wide range of soils and conditions. They are abundant on roadsides, drainage ditches, ranchlands, marshes, and wet grasslands and can be found sneaking into agricultural fields that are neglected and into gardens where the ground is disturbed.

Medicinal Uses

Parts used: Whole plant

TCM: Da ji (*Herba seu Radix Cirsii japonicum*), "big thistle," whole plant; sweet, bitter, cool; associated with the liver, heart

> 1. Cools blood, stops bleeding
> 2. Disperses blood stasis, reduces swelling, promotes healing
> 3. Treats jaundice
> 4. Lowers blood pressure

Classically, thistle is used for various types of bleeding, toxic swellings, sores, and for abscesses, hypertension, and jaundice.

Ayurveda: Ayurvedic medicine uses thistles in similar ways as TCM.

Western botanical: Canada thistle leaf has been used for tuberculosis, skin eruptions, skin ulcers, and poison ivy rash. The root is used for dysentery, diarrhea, and worms. Santa Fe thistle (*Cirsium ochrocentrum*) was reportedly used by the Zuni Indians of the Southwest as a cure for syphilis. Hildegard of Bingen of the twelfth century describes how to use thistle: "A person who has eaten or drunk poison should pulverize the top, root, and leaves of thistle. He should consume this powder either in food or drink, and it will expel the poison. If someone has a rash on his body, he should mix this powder with fresh fat and anoint himself with it, and he will be healed."[40]

Plant Chemistry

Japanese thistle (*C. japonicum*): cirsimarin, pectolinarin, taraxasteryl acetate, stigmasterol, A-amyrin, B-amyrin, B-sitosterol, luleclin-7-glucoside, acacetin rhamnoglucoside, vitamin K

Canada thistle (*C. arvense*): *p*-coumaric acid, caffeic acid, ferulic acid, caftaric acid, chicoric acid, diferuloyltartaric acid, tiliacin

Pharmacological Actions

Antimicrobial, antioxidant, hemostatic, cardiovascular

Scientific Studies

Antimicrobial: In vitro tests showed the tincture of *C. japonicum* possessed inhibitory effects on *Mycobacterium tuberculosis*.

Antioxidant: A number of studies have been performed on various *Cirsium* species in relation to antioxidant activity. *Cirsium setidens* extract was found to protect the liver of rats who were induced with liver injury. It was concluded that the hepatoprotective action of *C. setidens* was related to the antioxidant compounds the plant possesses. As well, in vitro and animal studies showed *C. arisanense* roots and leaves/stems exhibit liver-protecting actions on par with silymarin, the compound from milk thistle. Not only did these extracts protect from liver toxicity, they also decreased the expression of hepatitis B surface antigen.

Anticancer: Two flavones (pectolinarin and 5, 7-dihydroxy-6, 4'-dimethoxyflavone) were isolated from *C. japonicum* to be analyzed for anticancer activity. In separate studies, these compounds were found to inhibit cancer cell and tumor growth, improve immune response, and increase life expectancy of mice.

Other Uses

Thistles can eaten after the spines are removed and remaining plant is cooked. See Rosemary Gladstar's story in the introduction to this plant, (page 290). The plant has also been used as a favorable food for cattle and horses after the leaves have been beaten or crushed in a mill to remove the prickles.

Ecological Importance

Thistle was cited in a study involving phytoremediation work with the explosive HMX on an antitank firing range. Bull thistle was collected from the contaminated site and was compared to other plants that were analyzed for remediation work. The pattern of HMX accumulation for the plants growing in a controlled environment was similar to that observed for the wild plants growing on the range. Therefore, the logical conclusion was that the wild plants and bull thistle grow in a specific place for a reason; they have taken root in the contaminated areas to address the toxicities, performing in ways that scientists are meticulously studying in the lab. In a separate study, *Cirsium comgestum* was found to be a hyperaccumulator of copper and a low to moderate accumulator of iron, zinc, and manganese.

Thistle is an abundant source of pollen and nectar for a wide variety of bees and insects.

> *How many insects a single one attracts! While you sit by it bee after bee will visit it, and busy himself probing for honey and loading himself with pollen, regardless of your overshadowing presence.*
>
> HENRY DAVID THOREAU, 1851

Harvesting and Preparing

Harvest and prepare as usual, fresh or dried.

Dosage

Medium- to high-dose botanical.

Decoction: The TCM dosage is 10 to 15 grams, up to 30 grams for dried, and up to 60 grams for fresh thistle.
Tincture: 30 to 60 drops.

Cautions and contraindications: Use with caution during pregnancy.
Herb and drug interactions: None known.

References

C. A. Groom, A. Halasz, L. Paquet, et al., "Accumulation of HMX (octahydro-1,3,5,7-tetranitri-1,3,5,7-tetrazocine) in Indigenous and Agricultural Plants Grown in HMX-contaminated Anti-tank Firing-range Soil," *Environmental Science and Technology* 36, no. 1 (January 2002): 112–18.

S. H. Lee, S. I. Heo, L. Li, et al., "Antioxidant and Hepatoprotective Activities of *Cirsium setidens* Nakai against CCl4-induced Liver Damage," *Am J Chin Med* 36, no. 1 (2008): 107–14.

J. Nazaruk, S. K. Czechowska, R. Markiewicz, et al., "Polyphenolic Compounds and In Vitro Antimicrobial and Antioxidant Activity of Aqueous Extracts from Leaves of Some *Cirsium* species," *Nat Prod Res* 22, no. 18 (December 2008): 1583–88.

K. L. Ku, C. T. Tsai, W. M. Chang, et al., "Hepatoprotective Effect of *Cirsium arisanense* Kitamura in Tacrine-treated Hepatoma Hep 3B cells and C57BL mice," *Am J Chin Med* 36, no. 2 (2008): 355–68.

S. Liu, J. Zhang, D. Li, W. Liu, et al., "Anticancer Activity and Quantitative Analysis of Flavone of *Cirsium japonicum* DC," *Nat Prod Res* 21, no. 10 (August 2007): 915–22.

S. Liu, X. Luo, D. Li, et al., "Tumor Inhibition and Improved Immunity in Mice Treated with Flavone from *Cirsium japonicum* DC," *Int Immunopharmacol* 6, no. 9 (September 2006): 1387–93.

B. E. Malayeri, A. Chehregani, N. Yousefi, et al., "Identification of the Hyper Accumulator Plants in Copper and Iron Mine in Iran," *Pakistan Journal of Biological Sciences* 11, no. 3 (2008): 490–92.

Tree-of-Heaven
Ailanthus altissima

Common names: Ailanthus, Chinese sumac, stinking sumac, tree-of-hell

Family: Quassia family (Simaroubaceae)

Related species: *Ailanthus excelsa* (India)

States where tree-of-heaven
is considered invasive

Some people called it tree of heaven. No matter where its seed fell, it made a tree which struggled to reach the sky. It grew in boarded-up lots and out of neglected rubbish heaps and it was the only tree that grew out of cement. It grew lushly, but only in the tenements districts. You took a walk on a Sunday afternoon and came to a nice neighborhood, very refined. You saw a small one of these trees through the iron gate leading to someone's yard and you knew that soon that section of Brooklyn would get to be a tenement district. The tree knew. It came there first. Afterwards, poor foreigners seeped in and the quiet old brownstone houses were hacked up into flats, feather beds were pushed out on the window sills to air and the tree of heaven flourished. That was the kind of tree it was. It liked poor people.

BETTY SMITH,
A TREE GROWS IN BROOKLYN

The Plant

Tree-of-heaven is an expression of the city landscape, rich with meaning, exploding in abundance, and growing where no other tree will. *Ailanthus* provides potent medicine for the land and people and is a symbol of hope and endurance for poor, desolate, urban communities like that described in Betty Smith's classic novel *A Tree Grows in Brooklyn*. When we put together a full picture of this unique tree, it begins to make sense: Its function at one level lends us understanding of its function at another level. For example, references to it thriving in the "sludges" of infested city environments leads to examples of its use as medicine for thousands of years to clean the "sludges" of the body: the bowels, and its wide use for diarrhea, dysentery, malaria, and parasites. Growing in infected landscapes, it addresses infectious diseases, like marlaria, so often rampant in these areas as well. In *King's American Dispensatory,* Dr. H. L. True "considers the presence of these trees in malarial districts to have a strong action . . . in antagonizing those influences that produce 'intermittents'[fevers]."[41] It also provides a breath of fresh air to the polluted cities with its cleaning of CO_2 emissions and with its use as a medicine for lung complaints such as asthma, bronchitis, and tuberculosis. Not only is it a valuable land remediator and medicine, but also tree-of-heaven has the potential to remediate whole communities by providing a boost to inner city economics and by instilling a deep, wildlife force that renews and uplifts the deprived spirit.

> Tree of Heaven is one of the most powerful and generally underrated medicinal plants in the U.S., specific for intestinal infestations such as giardia (there is a reason it likes to grow near streams) and breathing problems such as asthma. It literally connects the breath of the self to the heavens.
>
> STEPHEN HARROD BUHNER

Description

Tree-of-heaven is an extremely fast-growing, deciduous tree that can reach 80 to 90 feet tall. It has a relatively smooth, gray bark, and the tree sprouts from the ground tall and straight with few branches, then, at the crown it is broad and spreading. The leaves resemble sumac and some nut trees and are 1 to 4 feet long and divided into twelve to forty lance-shaped leaflets laying opposite each other upon a straight stem. Unlike sumac leaflets with teeth along the margin, tree-of-heaven has just one to five teeth at the base of the leaflet. Tree-of-heaven leaflets are also said to have one to four small round glands on their undersides. Small, greenish yellow flowers bloom in large clusters at the end of branches spring to summer. They turn into separate flat seeds that are surrounded by a tan papery wing that is eventually dispersed on the wind. The fresh bark and leaves stink when crushed.

> *It would be considered beautiful except that there are too many of it.*
>
> BETTY SMITH, *A TREE GROWS IN BROOKLYN*

Collection and Habitat

Native to China and Taiwan, missionaries and garden enthuthists introduced the tree-of-heaven to Europe in 1751 and to the United States in 1784. Chinese immigrants brought it with them to the West Coast during the California gold rush for medicine and for its ability to grow in varied environments. Tree-of-heaven is considered a noxious weed throughout much of the world and grows in almost all habitats. It tolerates alkalinity, disease, drought, frost, heat, high pH, hydrogen fluoride, insects, low pH, pollution, poor soil, SO_2 (sulfur dioxide), and water logging, and it seems to survive at times without sun, water, or soil. It has spread outward from the cities by rail line and highway, flourishing in disrupted, polluted areas where no other plants will grow.

Medicinal Uses

Parts used: Stem, root bark, and leaves

TCM: Chun (Gen) Pi (Cortex *Ailanthus altissima*), bark; bitter, astringent, cold; associated with the large intestine, stomach, liver, and lungs

1. Clears heat, dries dampness, binds intestines
2. Stops bleeding
3. Kills parasites

Classically, tree-of-heaven is used to treat dysentery, chronic diarrhea (especially if there is blood or pus), uterine bleeding, leucorrhea, prolapse, roundworms, and fungal infection.

Ayurveda: Aralu, Katvanga (*Ailanthus excelsa*), bark, leaf

Classically, tree-of-heaven is used to treat lesions, diarrhea, menstrual disorders, nonhealing ulcers, and mouth infections

Western botanical: Used by eclectics in much the same way as for TCM and ayurveda. Some additional uses include asthma, bronchitis, dyspepsia, earache, wounds, swellings, skin eruptions, palpitations, obstinate hiccups, epilepsy, tonsillitis, sore throat, typhoid fever, muscular spasms, and it clears phlegm from the lungs. *Ailanthus* is a relative of quassia the tropical tree used in similar ways as medicine and well known for treating dysentery, and contains quassinoids. In addition, peoples of Australia and Africa have similar uses for tree-of-heaven.

Plant Chemistry

Ailanthin, ailanthinone, ailanthone, amarolide, azelaic acid, B-sitosterol, calcium-oxalate, canthin, carboline, elaidic acid, fisetin, gallic acid, isoquercetin, isoquercitrin, linuthin, malanthin, mersosin, neoquassin, oleoresin, quassine, neoquassine, nersosine, oleoresin, phlobaphene, quassin, quercetin, rhamnose, saponin, scopoletin, shinjulactone, syringic acid, tannin, vanillic acid, vitaxin

Pharmacological Actions

Astringent, febrifuge, anthelmintic, antispasmodic, antiseptic, expectorant, antineoplastic.

Scientific Studies

Anticancer: The quassinoids found in *Ailanthus* species are being extensively studied for their anticancer properties. In one study, the effects of one such component from *A. altissima* indicated that it was active in human leukemia, thyroid carcinoma, and liver carcinoma cell lines and was to be considered for further in vivo studies. In another study a compound isolated from the root bark of *A. excelsa* was comparable with the antitumor drugs paclitaxel and cisplatin in a variety of cancer cell lines in vitro and in mice studies.

HIV: The stem bark of *Ailanthus altissima* was found by Korean researchers to be the most potent species to inhibit fusion of the human immunodeficiency virus type 1 (HIV-1), with 74.9 percent +/- 4.4 percent at a concentration of 100 micrograms/milliliter.

Antimalarial/Antimicrobial: The component ailanthone is being studied for its antimalarial actions, and other antimicrobial constituents from fruits of the *Ailanthus* species appear to have in vitro activity against a broad spectrum of bacterial, viral, parasitic, and fungal infections.

Hematochezia: In a Chinese study, an herbal decoction using 120 grams of *Ailanthus* bark, three times daily, effectively helped all of 46 patients with blood in the stool due to various causes.

Anti-inflammatory: *Ailanthus altissima* was studied in vitro and in vivo and found to possess anti-inflammatory activity in ovalbumin-induced lung inflammation.

Antifertility: A couple of studies found *Ailanthus excelsa* extract in vitro to have a significant antiestrogenic effect, and it has long been considered a contraceptive agent by the people of India, where this species grows.

Other Uses

Tree-of-heaven is a renewable biomass energy and firewood source, growing up to 12 feet in height during a five-month growing season, and it readily coppices (resprouts from stump or from roots when cut down). The moderately hard and heavy wood is used for charcoal and firewood in many countries and makes a good candidate for energy production.

It has been cultivated for silk production—although it was said to possess an inferior gloss and texture and is also unable to take dye. This fabric, however, is perfect for hankies and underwear. Tree-of-heaven seems to be a reasonable choice to boost inner city economics, where many welfare-bound individuals could reap the rewards of the ancient practice of silk making by connecting to the natural world around them and using resources right outside their doorstep with little up-front investment.

The wood is used for sculpture and construction, and in India, the resin provides an incense used in Hindu temples.

Ecological Importance

The pervasive influence of tree-of-heaven on city life is due to its importance in cleaning the carbon dioxide emissions and air pollutants that are so commonplace in the urban landscape. There also appears to be evidence of its ability to clean other heavy-metal influences in the soil and water, for it thrives where few other plants can even take root. In preliminary in vitro testing for phytoremediation work, ailanthus exposed to the heavy metals copper, zinc, and magnesium demonstrated a tolerance comparable to species already utilized in the field. The plant is now being explored for remediation of a wide range of toxins due to the tree's high tolerance for a variety of extreme conditions and contaminants.

Harvesting and Preparing

Exposure to the sap can cause skin irritations for some, and pose a serious problem for others, due to its influence on the heart muscle. According to TCM, the root bark is considered to be of better quality than the stem bark. The best-quality bark is cut during the spring. Charring the root bark enhances its action to bind intestines for diarrhea. It can also be prepared externally as a wash or ointment, or used internally as an enema.

Dosage

Low- to medium-dose botanical. Combine with other antiparasitic herbs to lessen digestive upset.

Decoction: The TCM dosage is 3 to 5 grams.

Tincture: 5 to 20 drops.

Cautions and contraindications: Use this plant with caution. Internally, large doses may cause nausea and vomiting and may make a patient incoherent, sleepy, and weak. See "Harvesting and Preparing" for information on handling the plant's sap.

Herb and drug interactions: None found. Due to its cardiac influence, use caution with heart medications.

References

S. Dhanasekaran, B. Suresh, M. Sethuraman, et al., "Antifertility Activity of *Ailanthus excelsa* Linn. in Female Albino Rats," *Indian J Exp Biol* 31, no. 4 (1993): 384–85.

V. Ravichandran, B. Suresh, M. N. Sathishkumar, et al., "Antifertility Activity of Hydroalcoholic Extract of *Ailanthus excelsa* (Roxb): An Ethnomedicine Used by Tribals of Nilgiris Region in Tamilnadu," *J Ethnopharmacol* 112, no. 1 (2007): 189–91.

S. Rahman, N. Fukamiya, M. Okano, et al., "Anti-tuberculosis Activity of Quassinoids," *Chem Pharm Bull* (Tokyo) 45, no. 9 (1997): 1527–29.

M. Shrimali, D. C. Jain, M. P. Darokar, et al., "Antibacterial Activity of *Ailanthus excelsa* (Roxb). *Phytother Res* 15, no. 2 (2001): 165–66.

C. C. Zhao, J. H. Shao, X. Li, et al., "Antimicrobial Constituents from Fruits of *Ailanthus altissima* SWINGLE," *Arch Pharm Res* 28, no. 10 (2005): 1147–51.

E. Gatti, "Micropropagation of *Ailanthus altissima* and *In Vitro* Heavy Metal Tolerance," *Biologia Plantarum* 52, no. 1 (March 2008).

M. Ogura, G. A. Cordell, A. D. Kinghorn, et al., "Potential Anticancer Agents vs. Constituents of *Ailanthus excelsa* (Simaroubaceae)," *Lloydia* 40, no. 6 (1997): 579–84.

Y. S. Chang and E. R. Woo, "Korean Medicinal Plants Inhibiting to Human Immunodeficiency Virus Type 1 (HIV-1) Fusion," *Phytother Res* (4) (April 2003): 426–29.

A. L. Okunade, R. E. Bikoff, S. J. Casper, et al., "Antiplasmodial Activity of Extracts and Quassinoids Isolated from Seedlings of *Ailanthus altissima* (Simaroubaceae)," *Phytother Res* 17, no. 6 (2003): 675–77.

J. Polonsky, Z. Varon, C. Moretti, et al., "The Antineoplastic Quassinoids of *Simaba cuspidata* Spruce and *Ailanthus grandis* Prain, *J Nat Prod* 43, no. 4 (1980): 503–9.

A. A. Seida, A. D. Kinghorn, G. A. Cordell, et al., "Potential Anticancer Agents IX. Isolation of a New Simaroubolide, 6 alpha-tigloyloxychaparrinone, from *Ailanthus integrifolia* ssp. Calycina (Simaroubaceae)," *Lloydia* 41, no. 6 (1978): 584–87.

V. De Feo, L. D. Martino, A. Santoro, et al., "Antiproliferative Effects of Tree-of-heaven (*Ailanthus altissima* Swingle)," *Phytother Res* 19, no. 3 (2005): 226–30.

L. A. Anderson, A. Harris, and J. D. Phillipson, "Production of Cytotoxic Canthin-6-one Alkaloids by *Ailanthus altissima* Plant Cell Cultures," *J Nat Prod* 46, no. 3 (1983): 374–78.

M. S. Lavhale, S. Kumar, S. H. Mishra, et al., "A Novel Triterpenoid Isolated from the Root Bark of *Ailanthus excelsa* Roxb (Tree of Heaven), AECHL-1 as a Potential Anti-cancer Agent," *PLoS ONE.* 4, no. 4 (2009): e5365.

White Mulberry
Morus alba

Common names: white mulberry, Chinese mulberry

Family: Mulberry family (Moraceae)

Related species: red mulberry (*M. rubra*), black mulberry (*M. nigra*), Texas mulberry (*M. microphylla*), paper mulberry (*Broussonetia papyrifera*)

States where white mulberry is considered invasive

The Plant

Initially, white mulberry was not on my list of plants to be highlighted in this section. I included it only after continually finding *Morus alba* referenced as containing numerous potent compounds that address pathogenic influences and toxic accumulations. Various parts of the mulberry tree have been used in TCM for thousands of years, and with their cooling, detoxifying, and nurturing virtues, each has an impact on a different bodily system. The leaves mainly affect the upper part of the body, the eyes, head, throat, and lungs, helping to clear infection and inflammation. The twigs influence circulation to the joints and limbs to relieve arthritis and muscle stiffness. The root bark and bark of the tree work on the lower body, promoting urination and reducing edema, and the fruit nourishes the blood and hence the whole body but has a specific influence on the heart.

Then there's fears of hybridizing with the red mulberry.

Hybridization is a natural process of all species of life that helps plants, trees, animals (yes, humans too) to adapt to pressures from the environment. Consolidating resources and genes provides greater resistance to disease, toxins, and other unforeseen forces. The American elm was devastated by disease due to the fact that the tree's genetic immune system hadn't developed resistance to it. Now scientists hybridize the Siberian elm with the American elm to foster a tree that is highly adapted and resistant to Dutch elm disease, and the crossing of strains will help the elm tree to better survive. If there happens to be a mulberry blight, this genetic variation between the red and white species would most likely aid future generations of the tree.

Description

White mulberry trees are easily recognized by their coarsely toothed, heart-shaped or 3 to 5 lobed leaves that are smooth and glossy and approximately 2 to 8 inches long. (Red mulberry leaves are generally rougher looking, and paper mulberry has fuzzy leaves.) They exude a milky sap from the leaf stems. The bark on white mulberry is light gray with ridges running along it, and the tree grows 20 to 60 feet with spreading branches. Nondescript flower clusters form blackberry-like fruits in early to mid summertime that are whitish to purple in color.

Collection and Habitat

White mulberry originates from China and was brought to the United States by the British in the 1700s to establish a silk industry. Though the silk worms miserably failed, the tree did not and has since been planted and widely naturalized. White mulberry now flourishes throughout North America and often grows along field edges and roadsides and enters meadows and floodplains.

Medicinal Uses

Parts used: leaves, twig, bark, and fruit
TCM: Sang ye (*Folium Mori alba*), leaf; sweet, bitter, cold; associated with the lungs and liver

1. Clears wind heat
2. Calms the liver and clears the eyes
3. Clears the lungs and moistens dryness
4. Cools blood and stops bleeding

Classically, white mulberry leaf is used to treat fever, headache, sore throat, red, sore, dry or painful eyes, eye floaters, vertigo, dry cough, and cough with thick, yellow sputum.

Sang zhi (*Ramulus Mori alba*), twig; bitter, neutral; associated with the liver

1. Benefits the joints and promotes circulation

Classically, white mulberry twig is used to treat joint pain, arthritis, edema, numbness, and itching, especially in upper limbs, and in hemiplegia as sequela to a stroke.

Sang bai pi (*Cortex Mori alba*), root bark, bark; sweet, cold; associated with the lungs

1. Clears heat from lungs, stops cough, and calms wheezing
2. Promotes urination and reduces edema
3. Used for hypertension

Classically, white mulberry bark is used to treat cough with fever, irritability, and thirst, wheezing, edema, facial/eye swelling, and urinary difficulty.

Sang shen (*Fructus Mori alba*), fruit; sweet, cold; associated with heart, liver, kidney

1. Tonifies the blood and enriches yin
2. Moistens intestines for constipation

Classically, white mulberry fruit is used to treat dizziness, tinnitus, insomnia, premature aging, and constipation.

Ayurveda: No known uses found.

Western botanical: The eclectics described red mulberry fruit as being nutritious, with cooling and laxative actions, specific for febrile diseases with much thirst. The bark was also reputed to be purgative and a vermifuge used for tapeworms.

Plant Chemistry

Acetic acid, acetone, adenine, albafuran, albanin, albanol, amyrin, aluminum, arachidonic acid, arsenic, ascorbic acid, ash, asparagic acid, astragalin, benzaldehyde, beta-carotene, B-ecdysterone, B-sitosterol, betulinic acid, biotin, butylamine, caffeic acid, calcium, campesterol, caproic acid, chalcomoracin chlorogenic acid, choline, chrysanthemin, citric acid, copper, cudranin, cyaniding, cyclomorusin, cyclomuberrin, ecdysterone, eugenol, folacin, folinic acid, fructose, GABA, glucose, glutamic acid, glutathione, guaiacol, hemolysin, hexaldehyde, inokosterone, iron, isobutyric acid, isoquercetin, isoquercitrin, kaempferol, kuwanon, leucine, linalool, linoleic acid, lupeol, m-cresol, maclurin, magnesium, malic acid, manganese, moracenin, moracetin, morin, morusin, mulberranol, mulberrin, mulberrochromene, muberrofuran, myristic acid, n-butanol, n-butyraldehyde, niacin, norartocarpanone, o-cresol, oleic acid, oxalic acid, oxydiydromorusin, p-cresol, palmitoleic acid, pentanoic acid, phosphorus, piceatannol, pipecolic acid, potassium, proline, propionic acid, quercetin, quecitrin, resorcinol, resveratrol, riboflavin, rutin, sarcosine, scopoletin, scopolin, silica, sitosterol, sodium, stearic acid, succinic acid, sulfur, tannin, tartaric acid, thiamin, trigonelline, umbelliferone, valeraldehyde, valeric acid, xanthophylls, and zinc

Pharmacological Actions

Antimicrobial, antioxidant

Scientific Studies

Many of the potent compounds found in white mulberry have individually been studied numerous times to verify antioxidant, antimicrobial, liver-protecting, and health-promoting benefits.

Antimicrobial: The leaf has in vitro shown inhibitory effects against *Staphylococcus aureus,* B-hemolytic streptococcus, *Bacillus diphtheria, E. coli, Pseudomonas aeruginosa,* and leptospirosis. White mulberry leaf was also found to possess potent antisnake venom actions, and helped protect against local and systemic effects of viper (*Daboia russelii*) poison.

Antioxidant: White mulberry contains numerous antioxidant compounds including resveratrol, quercetin, rutin, vitamins A, B, C, trace minerals, sulfur, and other mulberry-specific constituents, all of which could be applied to numerous ailments. Studies on mulberry leaf extract showed that long-term administration has antihyperglycemic, antioxidant, and antiglycation effects in chronic diabetic rats. The extract lowered blood glucose levels and protected rats from oxidative stress.

Anti-atherosclerosis: In another study to verify traditional use, white mulberry fruit has been found to help prevention of coronary heart disease. Both a water extract and an anthocyanin-rich extract from the fruit exhibited antioxidative and anti-atherosclerogensis actions in vitro. The extracts were shown to have a great ability to scavenge for free radicals; the concentrated anthocyanin extract displayed efficiency that was ten times greater than that of the water extract.

Anti-inflammatory: Several studies show the effects of the isolated compounds mulberroside A and oxyresveratrol, both from white mulberry bark. Both were verified as effective at reducing inflammation in vitro and in vivo.

Other studies: The leaf was reportedly effective in treatment of 1,638 patients with elephantiasis. After three treatment courses of fifteen to twenty days each, 99 percent showed improvement.

Other Uses

The fruit has been used as a food and the leaves as fodder for livestock. The Chinese have long employed white mulberry for the ancient art of silk production.

Ecological Importance

In vitro studies concurred with field observations that showed that white mulberry root exudate helps a microorganism (*Sphingomonas yanoikuyae* JAR02) remove benzoapyrene, a polycyclic aromatic hydrocarbon, from the environment.

Harvesting and Preparing

Harvest and prepare as usual throughout the year. The fruit can be prepared to make a medicinal syrup.

Dosage

Medium-dose botanical.

Decoction: The TCM dosage is 5 to 15 grams of the leaf, bark, or fruit and 10 to 30 grams of the twig.

Cautions and contraindications: Allergic skin reactions have been reported from consuming the fruit, occurring thirty minutes after ingestion with a rash that spreads over the body. The fruit should be consumed sparingly by those with diarrhea and weak digestion.

Herb and drug interactions: None known.

References

J. A. Rentz, P. J. Alvarez, and J. L. Schnoor, "Benzo[a]pyrene Co-metabolism in the Presence of Plant Root Extracts and Exudates: Implications for Phytoremediation," *Environ Pollut* 136, no. 3 (August 2005): 477–84.

J. Naowaboot, P. Pannangpetch, V. Kukongviriyapan, et al., "Antihyperglycemic, Antioxidant and Antiglycation Activities of Mulberry Leaf Extract in Streptozotocin-induced Chronic Diabetic Rats," *Plant Foods Hum Nutr* 64, no. 2 (June 2009): 116–21.

L. K. Liu, H. J. Lee, Y. W. Shih, et al., "Mulberry Anthocyanin Extracts Inhibit LDL Oxidation and Macrophage-derived Foam Cell Formation Induced by Oxidative LDL," *J Food Sci* 73, no. 6 (August 2008): 13–21.

K. T. Chandrashekara, S. Nagaraju, S. U. Nandini, et al., "Neutralization of Local and Systemic Toxicity of *Daboia russelii* Venom by *Morus alba* Plant Leaf Extract," *Phytother Res* 23, no. 8 (February 20, 2009): 1082–87.

K. O. Chung, B. Y. Kim, M. H. Lee, et al., "In-vitro and In-vivo Anti-inflammatory Effect of Oxyresveratrol from *Morus alba* L.," *J Pharm Pharmacol* 55, no. 12 (December 2003): 1695–700.

H. Oh, E. K. Ko, J. Y. Jun, et al., "Hepatoprotective and Free Radical Scavenging Activities of Prenylflavonoids, Coumarin, and Stilbene from *Morus alba*," *Planta Med* 68, no. 10 (October 2002): 932–34.

S. Y. Kim, J. J. Gao, W. C. Lee, et al., "Antioxidative Flavonoids from the Leaves of *Morus alba*," *Arch Pharm Res* 22, no. 1 (February 1999): 81–85.

States where wild mustard
is considered invasive

Wild Mustard
Brassica rapa var. *nigra*

Common names: wild mustard, field mustard
(*B. rapa*), black mustard (*B. nigra*)

Family: Mustard family (Brassicaceae)

Related species: There are more than a hundred species in the Brassica family including some that are widespread: white mustard (*B. alba*), rapeseed (*B. napus*), and India mustard (*B. juncea*)

> *The kingdom of heaven is like to a grain of mustard seed, which a man took, and sowed in his field: Which indeed is the least of all seeds: but when it is grown, it is the greatest among herbs.*
>
> MATTHEW 13

The Plant

Mustard invigorates those places that are congealed, fixed, rocklike, as it does for the baked soils and gravelly terrain where it persists. Mustard is a hot, energizing, and circulating medicine that is used to treat rigid muscles, arthritic joints, congealed phlegm, nodules, and tumors. Wild mustard, just like many of the cultivated Brassicas (cabbage, cauliflower, broccoli, collards, kale, turnips, and so forth), contains a variety of vitamins, minerals, and potent compounds, many of which are cited by the National Cancer Institute as having anticancer effects. A mustard paste on the upper back relieves asthma, bronchitis, pneumonia, pleurisy, sinusitis, and other complaints of the lungs and upper respiratory system. The medicine dries the mucous membranes of the body, helping to remove phlegm from the respiratory and digestive system. The tissues are quickly dried by virtue of the plant's heating quality that penetrates deeply and removes excess growths and mucous buildups. This is also why we must be careful with its application: it can burn exposed tissues that are not covered by sufficient mucus. As in the environment, however, fire can have restorative properties for the soil and life underground. Likewise, some people with nerve damage can benefit from mustard's heating stimulation. This can help wake up the nerve endings and stimulate impulses in order to advance the healing of the nerves. Mustard restores vitality with its invigorating and stimulating attributes; it provides a much needed fire in those who tend to be sluggish and fixed and helps to aerate the barren and irritated terrain, both within and without.

Description

Wild mustard is an annual herb growing 2 to 3 feet tall and blooming from early summer to late fall, with abundant bright yellow, four-petaled flowers. Long slender pods form and stand erect as the seeds ripen. In black mustard (*B. nigra*), the pods hug the stem, whereas in field mustard (*B. rapa*), the pods stand off the stem. The gray-green leaves of field mustard are sparsely toothed or divided, and in the upper leaves they clasp to the stem and have earlike lobes. The black mustard leaves are bristly and coarsely lobed with upper leaves that are lance shaped and have no hairs.

Collection and Habitat

Wild mustards have no boundaries and have been spreading far and wide for a long time. Various mustard species were brought over with the earliest European settlers through contaminated feed and seed, and now they occupy a niche throughout all regions of North America. They can be found in waste places and gravelly roadsides, and it is considered a nuisance by farmers and gardeners alike, as it attempts to reclaim the disturbed soil that tractors and hoes continually create.

Medicinal Uses

Parts used: Leaf, seed

TCM: Bai jie zi (*Brassica juncea*), seed; acrid, warm; associated with the lungs

1. Warms the lungs, expels phlegm
2. Promotes circulation, alleviates pain

Classically, white mustard was used to treat cough with copious and thin sputum; numbness, stiffness, and pain in extremeties, abscesses, sores, nodules, and scrofula

Ayurveda: Raajikaa, Aasuri Raai, seed, oil

Classically, white mustard was used to treat intestinal catarrh, colic pain, indigestion, flatulence, rhinitis, arthritis, enlarged liver, spleen, and internal abscesses. Externally, the oil has been used for itching, inflammations, skin diseases, muscular rigidity, scrofula, and fistula.

Western botanical: Mustard has been employed since ancient times for numerous ailments. Nicholas Culpeper wrote about yellow mustard:

> It is excellent for such whose blood wants clarifying . . . for weak stomachs. It strengthens the heart and resists poison. The decoction of the seeds made in wine and drank provokes urine. The seed helps the spleen and pain in the sides, and gnawings in the bowels, and used as a gargle for the diseases of the throat. The outward application hereof upon the pained place of the sciatica eases the pains, as also the gout and other joint aches. The seeds bruised mixed with honey and applied or made up with wax, takes away the marks and black and blue spots of bruises, freckles or the like, the roughness or scabbiness of the skin, as also the leprosy.[42]

Plant Chemistry

Allyl-cyanide, allyl-isothiocyanate, allyl-rhodanide, arginine, ascorbic acid, ash, beta-carotene, caffeic acid, calcium, chlorogenic acid, copper, cystine, eo, erucic acid, ferulic acid, gadoleic acid, histidine, hydroxybenzoic acid, iron, isoleucine, leucine, linoleic acid, lysine, magnesium, manganese, mehionine, mucilage, myrosin, niacin, oleic acid, p-coumaric acid, palmitic acid, palmitoleic acid, pantothenic acid, phenylalanine, phenylethyl-isothiocyanate, phosphorus, potassium, protocatechuic acid, riboflavin, sinapic acid, sinapin, sinigrin, sodium, stearic acid, thiamin, threonine, trans-cinnamic acid, tryptophan, tyrosine, valine, vanillic acid, and zinc

Pharmacological Actions

Expectorant, anti-inflammatory, antimicrobial, antioxidant

Scientific Studies

Antimicrobial: In a study to determine the antibacterial effects in preserving food, scientists found yellow mustard powder to be an important ingredient inhibiting *E. coli* from surviving in sausage.

Anti-inflammatory: In one study, fifty children with acute or chronic tracheitis who were treated with mustard seed (Bai jie zi) had good results. A paste was made of mustard powder mixed with flour (1:3) and water and applied to the upper back in the evening. It was removed in the morning. There was a total of two or three treatments.

Antihyperglycemic: *Brassica nigra* seed was found to have antidiabetic activity in preliminary studies conducted on rats. Over the course of one month, a once-daily water extract of mustard seed brought down fasting serum glucose (FSG) levels and moderated the levels of glycosylated hemoglobin (HbA1c) and serum lipids better than the untreated control group.

Other studies: In a clinical study of 1,052 patients suffering from numbness of the face, mustard paste (Bai jie zi) was applied to affected areas and was covered with gauze for three to twelve hours. If necessary, the treatment was repeated in ten to fourteen days. Of the original 1,052 patients, 915 remained throughout the study, and there was a 97.7 percent effectiveness rate among them.

Other Uses

Wild mustard is related to cultivated mustards and brassicas, and the leaf can be added to salads and stir-fries for the spicy mustard flavor that has more bitterness than the cultivars. The seeds are also a source of wild flavor and can be ground and used in place of commercial mustard. In addition, I learned, while paging through the scientific studies on mustard, that the seed can be used as a potent preservative in foods; it appears to protect against *E. coli* and possibly a wide variety of food-borne pathogens.

Mustard is an important source of pollen and nectar for a variety of bees and insects, especially where other wild flowers do not grow. As a biofuel, mustard is a relative of rapeseed (*Brassica napus*), from which the oils are extracted from the seeds and converted into a power source.

Ecological Importance

Mustard produces powerful sulfur-rich compounds and phytochelatins that help account for it ability to tolerate diverse soil situations and heavy-metal contaminations. Indian mustard (*B. juncea*) has been found in laboratory and field studies to have the potential to remediate nickel (Ni), zinc (Zn), cadmium (Cd), chromium (Cr), and mercury (Hg) from toxic sites. The studies show its potential for phytofiltration of mercury-contaminated water and phytostabilization of contaminated soils. Also, with its sulfur compounds, the plant can detoxify cadmium in the environment. Scientists are genetically engineering Indian mustard for greater tolerance and accumulation capacity for these metals as well as for selenium (Se) and arsenic (As).

Harvesting and Preparing

Harvest and prepare the aerial parts of the plant throughout the year. To prepare mustard paste, grind seeds into a powder, then mix in a small amount of hot water or vinegar to form a thick and moist paste. You can combine the paste with more flour to dilute the mustard. Apply to the skin and cover with a bandage. See "Cautions and Contraindications," below.

Dosage

Medium-dose botanical.

Decoction: TCM dose is 3 to 9 grams and up to 15 grams with strong constitutions. The seed should not be decocted for long in tea preparations: approximately five minutes. The acrid elements are volatile and must by released by either water or heat, though overcooking will disperse them into the air.

Cautions and contraindications: There are reports of allergic reactions—both internally and externally—to wild mustard. It can irritate the skin with topical application, therefore start with small amounts to determine reaction levels. The paste should not be left on the skin for more than 15 to 20 minutes and should be kept away from the eyes. Use caution on those with sensitive skin. The heating quality can irritate the mucosa of the gastrointestinal tract as well, so use with caution internally when treating ulcers. Mustard can cause diarrhea, enteritis, and abdominal pain, but lowering the dose should relieve these issues.

Herb and drug interactions: None known.

References

C. A. Groom, A. Halasz, L. Paquet, et al., "Accumulation of HMX (octahydro-1,3,5,7-

tetranitri-1,3,5,7-tetrazocine) in Indigenous and Agricultural Plants Grown in HMX-contaminated Anti-tank Firing-range Soil," *Environmental Science and Technology* 30, no. 1 (January 2002): 112–18.

G. H. Graumann and R. A. Holley, "Inhibition of *Escherichia coli* O157:H7 in Ripening Dry Fermented Sausage by Ground Yellow Mustard," *Journal of Food Protection* 71, no. 3 (March 2008): 486–93.

P. Anand, K. Y. Murali, V. Tandon, et al., "Preliminary Studies on Antihyperglycemic Effect of Aqueous Extract of *Brassica nigra* (L.) Koch in Streptozotocin Induced Diabetic Rats," *Indian Journal of Experimental Biology* 45, no. 8 (August 2007): 696–701.

S. Reisinger, M. Schiavon, N. Terry, and E. A. Pilon-Smits, "Heavy Metal Tolerance and Accumulation in Indian Mustard (*Brassica juncea* L.) Expressing Bacterial Gamma-glutamylcysteine Synthetase or Glutathione Synthetase," *Int J Phytoremediation* 10, no. 5 (2008): 440–54.

S. Shiyab, J. Chen, F. X. Han, et al., "Phytotoxicity of Mercury in Indian Mustard (*Brassica juncea* L.)," *Ecotoxicol Environ Saf* 72, no. 2 (2009): 619–25.

G. Bañuelos, N. Terry, D. L. Leduc, et al., "Field Trial of Transgenic Indian Mustard Plants Shows Enhanced Phytoremediation of Selenium-contaminated Sediment," *Environ Sci Technol* 39, no. 6 (2005): 1771–77.

A. L. Wangeline, J. L. Burkhead, K. L. Hale, et al., "Overexpression of ATP Sulfurylase in Indian Mustard: Effects on Tolerance and Accumulation of Twelve Metals," *J Environ Qual* 33, no. 1 (2004): 54–60.

Wild Rose
Rosa spp.

Common names: Multiflora rose, rambler rose (*Rosa multiflora*), rugosa rose, Japanese rose, beach rose (*R. rugosa*)

Family: Rose family (Rosaceae)

Related species: There are many native and nonnative species of wild rose, as well as tens of thousands of dictict garden varieties (though they are often grafted onto wild rose rootstock for its vigor and disease resistance).

States where wild rose is considered invasive

The Plant

The prolific wild *Rosa* species has a bad reputation for colonizing everywhere, but it produces a lovely fragrance that has been longed for throughout history. Humans have had a constant love-hate relationship with rose.

> *You may want to wrap yourself up in her . . . until you realize she has thorns.*

Humans have used this plant as a food, medicine, ornament, and soil stabilizer for a long time, with the first wild rose cultivated a few thousand years ago in northern Persia. And through the course of history, with each country and culture it entered, stories of companionship have been told around this plant. Rose brings a loving caress to those who encounter it with open nose and eyes, and cultivars of the original wild roses are now widespread in gardens. This plant is so endeared to us as a species that the number of roses in gardens throughout the country could equal the number of wild roses in fields and on roadsides. None of the substantial stands of rose would be possible without the initial help of humans moving them around and planting them intentionally and then happily spreading with all the disturbances we create.

Oh, my, we could all talk about rose forever.
The petals of the rose are gentle sedatives, calming to the nerves,
and lift the spirits. Their beauty and strength are powerful
medicine for the "sick of heart" and they have their own built in
protection, which is what they offer us; soft power.

ROSEMARY GLADSTAR

Description

Multiflora rose has long arching canes with thorns throughout, often growing into large thickets. Its flowers, which bloom in late spring to summer, are small (0.5–1 inch), white to pinkish, and five-petaled with classic rose fragrance. They mature into small red fruits (hips) and survive well into winter.

Rugosa rose is a densely prickled and spreading plant with dark pink to white flowers (2 inches) that have five or more whorling petals and yellow stamens. This plant flowers throughout the year and matures in the fall into nice-sized hips (1–1.5 inches) that are red in color.

Collection and Habitat

Multiflora rose was introduced to the United States in 1866 from Japan and was originally used as rootstock for ornamental roses. Then later in the 1930s, it was widely planted for erosion control, living fences, and wildlife food and cover. Now multiflora rose is found throughout the temperate regions of North America, avoiding the high mountains, deserts, and subtropics. It favors disturbed land, old agricultural fields and pasture, and gaps and edges of woodlands. It is continually spread by agricultural practices and birds.

Rugosa rose was brought from Far East Asia to Europe and America in the late 1700s. It was introduced to hybridize with other rose varieties for its resistence to disease and was planted along beaches for erosion control and its tolerance to sandy soil, drought, and salt spray. Rugosa rose has a vast presence along the East Coast sand dunes and has moved westward to occupy other sandy environments and old fields.

Medicinal Uses

Parts used: Fruit, flower

TCM: Jin ying zi (*Fructus Rosae laevigatae*), "golden cherry fruit"; sour, astringent, neutral; associated with bladder, kidneys, and large intestine

> 1. Stabilizes kidneys, retains essence and urine
> 2. Binds the intestines and stops diarrhea
> 3. Regulates qi and promotes circulation

Classically, rose was used to treat frequent urination, incontinence, spermatorrhea, vaginal discharge, other urinary disorders, uterine bleeding, chronic diarrhea, prolapsed rectum or uterus, to "soothe a restless fetus," and for rheumatic pains.

Ayurveda: Used as in other traditions.

Western botanical: There are no references for the wild rose varieties from Asia, though other species such as dog rose (*R. canina*), field rose (*R. arvensis*), and red rose (*R. gallica*) have been used medicinally throughout Europe and Asia for thousands of years. Rose water is a soothing and cooling application for chapped hands and face, as well as for lesions, inflammations, and skin sores. Rose water can also be mixed with honey to make a gargle for sores in the mouth and throat. The hips are astringent and nutritive and can be used similarly to the applications listed for TCM: for urinary complaints, for digestive issues like diarrhea and dysentery, for coughs and spitting of blood, and for strengthening the heart. The hips were historically used as a sweetening agent for medicinal candy, in combination with other herbs.

Plant Chemistry

Afzelin, ascorbic acid, astragalin, B-sitosterol, campesterol, carotene, catechin, cholesterol, gallic acid, glycoside, isoquercitrin, kaempferol, multiflorin, multinoside, quercetin, quercitrin, riboflavin, rosamultin, rutin, salicylic acid, scoparone, sucrose, tiliroside, and tormentic acid

Pharmacological Actions

Antimicrobial, anti-inflammatory, antihyperlipidemic, antioxidant, antirheumatic

Scientific Studies

Antibiotic: A decoction of *R. laevigatae* has an inhibitory effect on various influenza viruses and a limiting influence on *Staphylococcus aureus, E. coli,* and *Pseudomonas aeruginosa.*

Antioxidant: Rose hips contain large amounts of vitamin C (1,000–2,000 mg/100 g) and provide the supplement industry with extracted forms of this well-known and potent antioxidant. It also contains the antioxidants rutin, quercetin, catechin, and kaempferol, among others.

Other studies: In one clinical study of 203 patients with prolapsed uterus, an herbal decoction of *R. laevigatae* had an effectiveness rate of 76 percent, with complete recovery in sixteen patients and improvement in 138 patients. In another clinical study of 20 infants with diarrhea, a decoction of *R. laevigatae* provided complete recovery in 13 cases, improvement in 6, and no response in only 1 case.

Other Uses

As Rosemary Gladstar says, "The berries make great jams and jellies. And they are good dry. I like to dry with the seeds in, makes a better tea and the seeds hold medicine. But for jams and jellies, seeds need to be removed."[43] Also, in Russia and Sweden a kind of wine is made by fermenting the fruit. The petals as well can be made into jams, syrups, preserves, candy, and vinegar, or added fresh to salads. The petals are useful as medicinal food, and rose water infuses subtle flavoring as well as providing a base for oitments; it can also be an essential oil or potpourri for perfume. A rose hip seed oil can also be used in skin care products and makeup.

Ecological Importance

Wild rose protects exhausted land, especially old agricultural fields. It also provides numerous species of wildlife with food and shelter.

Harvesting and Preparing

Harvest the budding flowers just before they open or, on bigger varieties, allow them to fully bloom and collect just the petals. Gather the fruit when the hips are ripe and bright red. Dry and prepare as usual.

Dosage

Medium- to high-dose botanical.

Decoction: The TCM dosage is 6 to 18 grams.

Cautions and contraindications: Wild rose is a safe food-grade herb for all constitutions. Due to its astringent nature and ability to bind the intestines, however, in excess it could cause constipation. It should not be taken at times of fever.

Herb and drug interactions: None known.

References

J. P. Joublan, M. Berti, H. Serri, et al., "Wild Rose Germplasm Evaluation in Chile," J. Janick, ed., *Progress in New Crops* (Arlington, Va: ASHS Press, 1996), 584–88.

Epilogue

First they ignore you, then they laugh at you, then they fight you, then you win.

MOHANDAS GANDHI

This is my "Manifesto against the Machine," if you will; the philosophically unconscious contraption that grids its wheels into the impoverished highway of Nature, with the doctrine of staying within the lines.

Don't get out of line, you hear!

Some may complain about the harshness of this exposé, with my frank character, unlimited exposure, and obvious prejudices defending these plants . . .

especially this annoying voice . . .

But I will contend, the reasoning was to balance out the overwhelming extremism against these plants, with few keeping it in check, and saying now the things so many of us have wished to hear for a long time now.

Or, I could proclaim . . .

"The plants made me do it!"

Are we looking to be cultivators of what we believe a "native" landscape to be? Or are we looking for a sense of the reclamation of the "wild," and being present to Mother Nature revealing her constantly changing wardrobe of life and landscape?

Watching gardeners label their plants
I vow with all beings
to practice the old horticulture
and let plants identify me.

ROBERT AITKEN, *The Dragon Who Never Sleeps*

In our every deliberation, we must consider the impact of our decisions on the next seven generations.

FROM THE GREAT LAW OF THE IROQUOIS CONFEDERATION

Ode to Manzanita

You remember that time you met Manzanita,
In the full moon light of the desert,
The red rocks shimmered all around you,
And you danced in her arms.
She held you like no other,
For your path had been lonely,
She taught you to surrender
And then you'd never be alone.
You walk this trail before you,
Forever changed by the endearment,
Now knowing whenever there's a bend in the road,
A plant will be there to guide you.
That time long ago, when you met Manzanita
In the full moon light of the desert,
The red rocks shimmered all around you,
And you remembered the existence of Me.

Notes

PART I. WAGING WAR ON PLANTS

1. Sylvan Ramsey Kaufman and Wallace Kaufman, *Invasive Plants: A Guide to Identification, Impacts, and Control of Common North American Species* (Mechanicsburg, Pa.: Stackpole Books, 2007).

CHAPTER I. THE POLITICS OF PROLIFIC PLANTS

1. Kaufman and Kaufman, *Invasive Plants,* 13.
2. Ibid., 13.
3. United States Congress, 1974. Federal Noxious Weed Act, 2801–14.
4. Executive Order 13112, February 3, 1999, Invasive Species, *Federal Register* 64, no. 25 (February 8, 1999).
5. Dana Joel Gattuso, "Invasive Species: Animal, Vegetable or Political?" *National Center for Public Policy Research* 544 (August 2006), in Kaufman and Kaufman, *Invasive Plants,* 13.
6. Mark Derr, "Alien Species Often Fit In Fine, Some Scientists Contend," *New York Times,* September 4, 2001.
7. See www.invasivespeciesinfo.gov/council/wrkgrps.shtml (accessed March 1, 2010); www.doi.gov/news/08_News_Releases/080801a.html (accessed March 1, 2010); David Theodoropoulos, *Invasion Biology: Critique of a Pseudoscience* (Blythe, Calif.: Avvar Books, 2003), 141.
8. D. Pimentel, L. Lach, R. Zuniga, et al., "Environmental and Economic Costs Associated with Non-indigenous Species in the United States," *Bioscience* 50 (2000): 53–65.
9. Arthur O'Donnell, "Invasive Species: Closing the Door to Exotic Hitchhikers," *Land Letter,* November 30, 2006, www.invasivespeciesinfo.gov/docs/resources/landletter20061130.pdf (accessed March 1, 2010).
10. Emily B. Roberson, "Plant Science and Conservation Groups Ask Congress

to Add Plants to Legislation Protecting Wildlife From Climate Change," on Native Plant Conservation Campaign, www.plantsocieties.org/Equal_ Protection.htm (accessed March 1, 2010).

11. Ibid.

12. Dana Joel Gattuso, "Invasive Species: Animal, Vegetable or Political?"

13. Wallace Kaufman, "Invasion of Alien Species," *PERC Reports* 21, no. 4 (December 2003).

14. National Agricultural Library (USDA), "Laws and Regulations of Invasive Species," www.invasivespeciesinfo.gov/laws/execorder.shtml (accessed March 1, 2010).

15. David Theodoropoulos, *Invasion Biology: Critique of a Pseudoscience,* Introduction; Kathryn R. Hoffman, "Alien Invasion," *Time for Kids* 8, no. 4 (October 4, 2002). M. A. Patten and R. A. Erickson, "Conservation Value and Rankings of Exotic Species," *Conservation Biology* 15 (2001): 817–18.

CHAPTER 2. THE SCIENCE OF INVASIONS

1. David Theodoropoulos, *Invasion Biology: Critique of a Pseudoscience,* 6.

2. Dave Jacke and Eric Toensmeier, *Edible Forest Gardens* (White River Junction, Vt.: Chelsea Green Publishing, 2005).

3. Ibid., 121.

4. D. C. Schmitz and D. Simberloff, "Biological Invasions: A Growing Threat," *Issues in Science and Technology* 13, no. 4 (Summer 1997): 33–41, www.nap .edu/issues/13.4/schmit.htm (accessed March 1, 2010).

5. Charles S. Elton, *The Ecology of Invasions by Animals and Plants* (Chicago: University of Chicago Press, 1958), 115.

6. Ibid., 115.

7. Carl Zimmer, "Friendly Invaders," *New York Times,* September 8, 2008.

8. M. A. Davis, K. Thompson, and J. P. Grime, "Charles S. Elton and the Dissociation of Invasion Ecology from the Rest of Ecology," *Diversity and Distributions* 7 (2001): 97–102.

9. Arthur O'Donnell, "Invasive Species: Closing the Door to Exotic Hitchhikers."

10. Charles S. Elton, *The Ecology of Invasions by Animals and Plants.*

11. David Theodoropoulos, *Invasion Biology: Critique of a Pseudoscience.*

CHAPTER 3. NATURALLY NATIVE:
PLANTS ON THE MOVE

1. David Theodoropoulos, *Invasion Biology: Critique of a Pseudoscience,* 21.

2. Constance I. Millar and Linda B. Brubaker, "Climate Change and Paleoecology:

New Contexts for Restoration Ecology," in M. Palmer, D. Falk, and J. Zedler, eds., *Restoration Science* (Washington, D.C.: Island Press, 2006), 315–40; Also, for shift in ranges of spruce forests, the image was borrowed originally from G. Jackson et al., eds., "North America and Adjacent Oceans During the Last Deglaciation," *Geological Society of America* 3 (1987): 277–88.

3. David Theodoropoulos, *Invasion Biology: Critique of a Pseudoscience,* 15.

4. National Aeronautics and Space Administration, "Solar Variability: Striking a Balance with Climate Change," *ScienceDaily* (May 12, 2008), www.science-daily.com/releases/2008/05/080512120523.htm (accessed July 12, 2009).

5. D. Blumenthlal, R. A. Chimner, J. M. Welker, et al., "Increased Snow Facilitates Plant Invasion in Mixed-grass Prairie," *New Phytologist* 179, no. 2 (July 2008): 440–48.

6. W. S. Schuster, K. L. Griffin, H. Roth, et al., "Changes in Composition, Structure and Aboveground Biomass over Seventy-six Years (1930–2006) in the Black Rock Forest, Hudson Highlands, Southeastern New York State," *Tree Physiology* 4 (April 28, 2008): 537–49.

7. Tom Christopher, "Can Weeds Help Solve the Climate Crisis?" *New York Times,* June 29, 2008.

8. Ibid.

9. Ibid.

10. L. H. Ziska, "Evaluation of the Growth Response of Six Invasive Species to Past, Present and Future Atmospheric Carbon Dioxide," *Journal of Experimental Botany* 54 (2003): 395–404.

11. Toby Hemenway, *Gaia's Garden* (White River Junction, Vt.: Chelsea Green Publishing, 2000), 13.

12. Masanobu Fukuoka, *The Natural Way of Farming* (Madras, India: Bookventure, 1985).

13. V. A. Nuzzo, J. C. Maerz, and B. Blossey, "Earthworm Invasion as the Driving Force Behind Plant Invasion and Community Change in Northeastern North American Forests," *Conservation Biology* 18 (February 2009).

14. Toby Hemenway, "Native Plants: Restoring to an Idea," www.patternliteracy.com/nativeplantsres.html (accessed March 1, 2010).

CHAPTER 4. THE NATURE OF HARM, THE HARM TO NATURE

1. National Invasive Species Council (NISC), "Invasive Species Definition Clarification and Guidance White Paper." Submitted by the Definitions Subcommittee of the Invasive Species Advisory Committee (ISAC), approved by ISAC April 27, 2006.

2. Dana Joel Gattuso, "Invasive Species: Animal, Vegetable or Political?"

3. Ibid.

4. Statement of Michael Soukup, Associate Director for Natural Resource Stewardship and Science, National Park Service, Department of the Interior, before the Subcommittee on National Parks, U.S. Senate Committee on Energy and Natural Resources, August 9, 2005; "Invasive Species: Scientists Demand Action on Invasive Species," Union of Concerned Scientists, Cambridge Massachusetts, January 18, 2006 (last revised), see www.ucsusa.org/invasive_species/call-to-action-on-invasive-species.html (accessed March 1, 2010); Invasive Species Management, Program Plan: 2003–2007, NASA, Office of Earth Science, Applications Program, June 6, 2003.

5. National Aeronautics and Space Administration (NASA), "Success at Stennis—A New Image for the Water Hyacinth," http://technology.ssc.nasa.gov/suc_stennis_water.html (accessed November 23, 2008).

6. The United States National Arboretum (USNA), "Invasive Plants," www.usna.usda.gov/Gardens/invasives.html (accessed March 1, 2010).

7. Ibid.

8. United States Environmental Protection Agency (EPA), "Threats to Wetlands," September 2001, www.epa.gov/owow/wetlands/pdf/threats.pdf (accessed March 1, 2010).

9. Kathy Keville, "American Ginseng," in Rosemary Gladstar and Pamela Hirsch, eds. *Planting the Future* (Rochester, Vt.: Healing Arts Press, 2000).

10. Jessica Gurevitch and Dianna K. Padilla, "Are Invasive Species a Major Cause of Extinctions?" *Trends in Ecology and Evolution* 19, no. 9 (September 2004): 470–74.

11. Mark Sagoff, "Do Invasive Species Threaten the Environment?" *Journal of Agricultural and Environmental Ethics* 18 (2005): 215–36; www.propertyrightsresearch.org/2005/articles04/do_invasive_species_threaten_the.htm (accessed August 24, 2008).

12. Ibid.

13. Ibid.

14. G. Vermeij, "When Biotas Meet: Understanding Biotic Interchange," *Science* 253, no. 5024 (September 6, 1991).

15. A. MacDougall, "Did Native Americans Influence the Northward Migration of Plants during the Holocene?" *Journal of Biogeography* 30, no. 5 (May 2003), 633–47, in Tom Christopher, "Can Weeds Help Solve the Climate Crisis?"

16. M. Davis, "Biotic Globalization: Does Competition from Introduced Species Threaten Biodiversity?" *Bioscience* 53 (2003): 481–89.

17. Ibid.

18. Mark Sagoff, "Do Invasive Species Threaten the Environment?"

19. G. Vermeij, "An Agenda for Invasion Biology," *Biological Conservation* 7 (1996): 83–89.

20. M. A. Davis, "Biotic Globalization: Does Competition from Introduced Species Threaten Biodiversity?" 481; Carl Zimmer, "Friendly Invaders"; F. Sax Dov and Steven D. Gaines, "Colloquium Paper: Species Invasions and Extinction: The Future of Native Biodiversity on Islands," *Proceedings of the National Academy of Sciences* 105, Suppl. 1 (2008): 11490–97.

21. Plataforma SINC, "Native Plants Can Also Benefit from the Invasive Ones," *ScienceDaily,* May 21, 2008, www.sciencedaily.com/releases/2008/05/080516125934.htm (accessed April 15, 2009).

22. D. E. Pearson, "Invasive Plant Architecture Alters Trophic Interactions by Changing Predator Abundance and Behavior," *Oecologia* 159, no. 3 (March 2009): 549–58.

23. S. O. Duke, A. C. Blair, F. E. Dayan, R. D. Johnson, K. M. Meepagala, D. Cook, and J. Bajsa, "Is (-)-catechin a Novel Weapon of Spotted Knapweed (*Centaurea stoebe*)?" *Journal of Chemical Ecology* 35, no. 2 (Feb 2009): 141–53.

24. L. G. Perry, G. C. Thelen, W. M. Ridenour, et al., "Concentrations of the Allelochemical Catechin in *Centaurea maculosa* Soils," *Journal of Chemical Ecology* 33 (2007): 2337–44.

25. N. Tharayil, P. Bhowmik, P. Alpert, et al., "Dual Purpose Secondary Compounds: Phytotoxin of *Centaurea diffusa* Also Facilitates Nutrient Uptake," *New Phytologist* 181, no. 2 (January 2009): 424–34.

26. Carl Zimmer, "Friendly Invaders"; H. A. Mooney and E. E. Cleland, "The Evolutionary Impact of Invasive Species," *PNAS* 98, no. 10 (2001): 5446–51.

27. S. M. Louda, D. Kendall, J. Connor, et al., "Ecological Effects of an Insect Introduced for the Biological Control of Weeds," *Science* 277, no. 5329 (August 22, 1997): 1088–90.

CHAPTER 5. INVASIVE HERBICIDAL IMPACTS

1. C. Cox, "Ten Reasons Not to Use Pesticides," *Journal of Pesticide Reduction* 26 (2006): 10–12; Carey Gillam, "Biotech Crops Cause Big Jump in Pesticide Use: Report," Rueters. Nov. 17, 2009.

2. "Exposure to Rainbow Herbicides," www.allmilitary.com/board/viewtopic.php?id=1209 (accessed February 17, 2008).

3. R. L. Tominack, G. Y. Yang, W. J. Tsai, et al., "Taiwan National Poison Center Survey of Glyphosate-surfactant Herbicide Ingestions," *J Toxicol Clin Toxicol* 29, no. 1 (1991): 91–109.

4. Environmental Protection Agency, www.naturescountrystore.com/roundup/page3.html (accessed February 12, 2009).

5. D. Monroe, 1989 Letter to the National Campaign against the Misuse of Pesticides (NCAMP), www.naturescountrystore.com/roundup/page2.html (accessed February 12, 2009).

6. Natural Resources Defense Council, "EPA Won't Restrict Toxic Herbicide Atrazine, Despite Health Threat," January 1, 2004, www.nrdc.org/health/pesticides/natrazine.asp (accessed March 1, 2010).
7. Robert Sanders, "Pesticide Atrazine Can Turn Male Frogs into Females," *ScienceDaily,* www.sciencedaily.com/releases/2010/03/100301151927.htm (accessed March 1, 2010).

CHAPTER 6. THE ECONOMICS OF WEEDS

1. D. Pimentel, L. Lach, R. Zuniga, et al., "Environmental and Economic Costs of Non-indigenous Species in the United States," *BioScience* 50 (2000): 53–65.
2. M. L. Corn, E. H. Buck, J. Rawson, et al., "Harmful Non-native Species: Issues for Congress," CRS Report for Congress, U.S. Congressional Research Service—Resources, Science, and Industry Division, Washington, D.C., April 8, 1999.
3. B. Tokar, "Agribusiness, Biotechnology and War," Institute for Social Ecology Biotechnology Project, September 1, 2002, www.social-ecology.org/2002/09/agribusiness-biotechnology-and-war (accessed July 8, 2009).
4. Kaufman and Kaufman, *Invasive Plants,* 3.
5. National Wildlife Refuge Association, "Silent Invasion: A Call to Action from NWRA," Oct 2002, www.refugenet.org/new-pdf-files/Silent%20Invasion%20pdf.pdf (accessed March 23, 2009).
6. Sabine Vollmer, "US: Pesticide Data May Tell Why Bees Die," *News Observer,* August 22, 2008.
7. Ryan McGee, "Leafy, Invasive, Irritating Kudzu Could Be NASCAR's Fuel of the Future," *ESPN The Magazine,* July 10, 2008.
8. John Roach, "'Grass Gas' Shows Promise as Superefficient, Clean Fuel," *National Geographic News,* January 9, 2008.
9. Dana Joel Gattuso, "Invasive Species: Animal, Vegetable or Political?" *National Center for Public Policy Research* 544 (August 2006).
10. Don Comis, "Metal-Scavenging Plants to Cleanse the Soil," *Agricultural Research* (November 1995), www.ars.usda.gov/is/AR/archive/nov95/cleanse1195.htm?pf=1 (accessed November 24, 2008).
11. Ilva Raskin, *Phytoremediation of Toxic Metals: Using Plants to Clean Up the Environment* (New York: John Wiley and Sons, 2000); Jeanna R. Henry, "An Overview of the Phytoremediation of Lead and Mercury," in National Network of Environmental Management Studies (NNEMS) status report 55, 2000.

PART 2. THE INTELLIGENCE OF PLANTS

1. Dale Pendell, *Pharmako/poeai* (San Francisco: Mercury House, 1995), introduction.

CHAPTER 7. THE DEEP ECOLOGY OF INVASIVES

1. James Lovelock, *Healing Gaia* (New York: Harmony Books, 1991), 17.
2. Ibid., 22.
3. O. Naidenko, N. Leiba, R. Sharp, et al., "Bottled Water Contains Disinfection Byproducts, Fertilizer Residue, and Pain Medication," *Environmental Working Group* (October 2008), www.ewg.org/reports/bottledwater (accessed July 7, 2009).
4. Ibid.
5. Martha Mendoza, "Study: Range of Pharmaceuticals in Fish across US," Associated Press, March 25, 2009.
6. Rebecca J. Goetz, "Cleaning Up Petroleum Spills with Plants," *Purdue University News Service,* June 26, 2000, www.scienceblog.com/community/older/2000/D/200003598.html (accessed November 20, 2008).
7. Lena Ma, Kenneth Komar, Cong Tu, et al., "A Fern that Hyperaccumulates Arsenic," *Nature* 409 (February 1, 200), http://lqma.ifas.ufl.edu/PUBLICATION/Ma-01a.pdf (accessed November 20, 2008).
8. Don Comis, "Phytoremediation: Using Plants to Clean Up Soils," *Agricultural Research Magazine* (June 2000), www.ars.usda.gov/is/ar/archive/jun00/soil0600.htm (accessed July 27, 2008).
9. Ibid.
10. U.S. Environmental Protection Agency, OSWER Directive 9200.4-17P, "Use of Monitored Natural Attenuation at Superfund, RCRA Corrective Action, and Underground Storage Tank Sites, April 21, 1999, Office of Solid Waste and Emergency Response, 41.
11. Amy Steward, *Wicked Plants* (Chapel Hill, N.C.: Algonquin Books, 2009), 88.

CHAPTER 8. THE CHEMISTRY OF PLANT MEDICINE

1. W. Cherdshewasar, S. Sriwatcharakul, and S. Malaivijitnond, "Variance of Estrogenic Activity of the Phytoestrogen-rich Plant," *Maturitas* 61, no. 4 (December 20, 2008): 350–57.
2. Stephen Harrod Buhner, *The Lost Language of Plants* (White River Junction, Vt.: Chelsea Green Publishing), 141.
3. Randolph E. Schmid, "Ants, Plants Mutually Benefit Each Other," Associated Press, January 10, 2008.
4. T. E. Gómez Álvarez-Arenas, D. Sancho-Knapik, J. J. Peguero-Pina, and E. Gil-Pelegrín, "Noncontact and Noninvasive Study of Plant Leaves Using Air-coupled Ultrasound," *Applied Physics Letters* 95, no. 19 (2009): 193702; "Leaves Whisper Their Properties through Ultrasound," www.scienceblog.com/cms/

leaves-whisper-their-properties-through-ultrasound.html (accessed February 3, 2010).

5. Ari Rabinovitch, "Scientists 'Listen' to Plants to Find Water Pollution," Reuters, August 14, 2008.

6. Stephen Harrod Buhner, *The Secret Teachings of Plants*.

7. Quoted in *HerbalGram* 26 (1992): 50.

8. David Hoffman, *Medical Herbalism* (Rochester, Vt.: Healing Arts Press, 2004), 12.

9. Kenny Ausubel, ed., *Ecological Medicine* (San Francisco: Sierra Club Books, 2004), 112.

10. Ibid., 111–12

11. A. C. Bronstein, D. A. Spyker, L. R. Cantilena Jr., et al., 2008 Annual Report of the American Association of Poison Control Centers' National Poison Data System (NPDS): 26th Annual Report, *Clinical Toxicology* (2009): 47, 911–1084.

12. Kenny Ausubel, ed., *Ecological Medicine,* 109.

13. B. Starfield, "Is US Health Really the Best in the World?" *JAMA* 284, no. 4 (July 26, 2000): 483–85.

14. Ibid.

CHAPTER 9. USING INVASIVE PLANTS
TO TREAT INVASIVE DISEASES

1. Stephen Harrod Buhner, *Healing Lyme* (Randolph, Vt.: Raven Press, 2005), 109.

2. Stephen Harrod Buhner, *The Lost Language of Plants,* 117.

3. Margie Mason and Martha Mendoza, "First Case of Highly Drug-resistant TB Found in US," Associated Press, December 27, 2009.

4. Stephen Harrod Buhner, *Herbal Antibiotics* (Pownal, Vt.: Storey Books, 1999), 2.

5. Martha Mendoza and Margie Mason, "Solution to Killer Superbug Found in Norway," Associated Press, December 31, 2009.

6. Stephen Harrod Buhner, *The Lost Language of Plants,* 116; D. Gutierrez, "Hospitals Flush 250 Million Pounds of Expired Drugs Into Public Sewers Every Year," *Natural News,* February 10, 2009, www.naturalnews.com/025573_drugs_water_hospitals.html (accessed March 23, 2009); Jeff Donn, Martha Mendoza, and Justin Pritchard, "AP Impact: Tons of Released Drugs Taint US Water," Associated Press, April 19, 2009.

7. Stephen Harrod Buhner, *Herbal Antibiotics,* 61.

8. Stephen Harrod Buhner, *Healing Lyme.*

9. Wikipedia; National Biological Information Infrastructure (NBII), http://

westnilevirus.nbii.gov/index.html (accessed March 23, 2009); Subhuti Dharmananda, "West Nile Virus," August 2002, at www.itmonline.org/arts/westnile.htm (accessed February 11, 2010).

10. University of Wisconsin, Madison, "Waterborne Disease Risk Upped In Great Lakes," *ScienceDaily,* October 11, 2008, www.sciencedaily.com/releases/2008/10/081008150522.htm (accessed March 18, 2009).

11. National Institutes of Health (NIH), MedlinePlus, www.nlm.nih.gov/medlineplus/ency/article/002473.htm (accessed March 16, 2009).

12 Ibid.

13. I. Maugh, H. Thomas, and Marla Cone, "Lead Exposure in Children Linked to Violent Crime," *Los Angeles Times,* May 28, 2008.

14. U.S. Department of Health and Human Services, Agency of Toxic Substances and Disease Registry, www.atsdr.cdc.gov/tfacts46.html#bookmark04 (accessed March 16, 2009).

15. Hoffmann, *Medical Herbalism,* 109, 111.

16. Z. Tunalier, M. Koşar, E. Küpeli, et al., "Antioxidant, Anti-inflammatory, Antinociceptive Activities and Composition of *Lythrum salicaria* L. Extracts," *Journal of Ethnopharmacology* 110, no. 3 (April 4, 2007): 539–47.

17. Hass, Elson, *Staying Healthy with Nutrition,* 257.

18. Kim Irwin, "Fruits, Vegetables, Teas May Protect Smokers from Lung Cancer," *UCLA News,* May 29, 2008.

19. Ute Nöthlings, Suzanne P. Murphy, Lynne R. Wilkens, et al., "Flavonols and Pancreatic Cancer Risk," *American Journal of Epidemiology* 166, no. 8 (2007): 924–31.

PART 3. GUIDE TO INVASIVE PLANTS:
THEIR MEDICINE AND ECOLOGICAL IMPORTANCE

1. Stephen Harrod Buhner, *Herbal Antibiotics,* 61.

2. Ibid.

3. Maud Grieve, *A Modern Herbal* (New York: Dover Publications, 1971).

4. Ibid.

5. Personal e-mail.

6. Maud Grieve, *A Modern Herbal.*

7. Dave Jacke and Eric Toensmeier, *Edible Forest Gardens,* 536.

8. James A. Duke, "Phytomedicinal Forest Harvest in the United States," www.fao.org/docrep/W7261e/W7261e17.HTM (accessed March 18, 2009).

9. V. L. Rodgers, B. E. Wolfe, L. K. Werden, et al., "The Invasive Species *Alliaria petiolata* (Garlic Mustard) Increases Soil Nutrient Availability in Northern Hardwood-conifer Forests," *Oecologia* 157, no. 3 (September 15, 2008): 459–71.

10. Ibid.
11. Ibid.
12. Maud Grieve, *A Modern Herbal*.
13. Timothy Coffey, *The History and Folklore of North American Wildflowers* (New York: Facts On File, 1993), 80.
14. Japanese Knotweed Alliance, "What is Japanese Knotweed?" www.cabi.org/japaneseknotweedalliance/Default.aspx?site=139&page=52 (accessed January 29, 2008). John Bailey, "Research on Japanese Knotweed," www.t-c-m-rd.co.uk/cms_misc/articles/Japanese_Knotweed_by_Dr_J._Bailey.pdf (accessed January 29, 2008).
15. John Bailey, "Research on Japanese Knotweed," www.t-c-m-rd.co.uk/cms_misc/articles/Japanese_Knotweed_by_Dr_J._Bailey.pdf (accessed March 19, 2010).
16. Paul Okunieff, Weimin Sun, Wei Wang, et al., "Anti-cancer Effect of Resveratrol Is Associated with Induction of Apoptosis via a Mitochondrial Pathway Alignment," *Advances in Experimental Medicine and Biology* 614 (2008): 179–86.
17. Maud Grieve, *A Modern Herbal*.
18. Ibid.
19. Timothy Coffey, *The History and Folklore of North American Wildflowers*.
20. Nicholas Culpeper, *The Complete Herbal* (London: Cumberland House, reprint 1995; original, 1653).
21. Sabrina Moreta, T. Populina, L. S. Contea, et al., "HPLC Determination of Free Nitrogenous Compounds of *Centaurea solstitialis* (Asteraceae), the Cause of Equine Nigropallidal Encephalomalacia," *Toxicon* 46, no. 6 (November 2005): 651–57.
22. L. Wu, H. Qiao, Y. Li, et al., "Protective Roles of Puerarin and Danshensu on Acute Ischemic Myocardial Injury in Rats," *Phytomedicine* 14, no. 10 (October 2007): 652–58.
23. L. C. Chiang, W. Chiang, M. Y. Chang, et al., "In Vitro Cytotoxic, Antiviral and Immunomodulatory Effects of *Plantago major* and *Plantago asiatica*," *American Journal of Chinese Medicine* 31, no. 2 (2003): 225–34 (edited slightly for readability).
24. François Couplan, *The Encyclopedia of Edible Plants of North America* (New Canaan, Conn.: Keats Publishing, 1998).
25. Personal e-mail.
26. Maud Grieve, *A Modern Herbal*, 498.
27. Ibid.
28. Personal e-mail.
29. Daniel Q. Thompson, R. L. Stuckey, E. B. Thompson, "Spread, Impact, and Control of Purple Loosestrife (*Lythrum salicaria*) in North American Wetlands," United States Department of the Interior, Fish and Wildlife.

30. Ibid.

31. Personal e-mail.

32. Maud Grieve, *A Modern Herbal.*

33. Ibid.

34. R. Sundararajan, N. A. Haja, K. Venkatesan, et al., *"Cytisus scoparius* Link—a Natural Antioxidant," *BMC Complement Altern Med* 6 (March 16, 2006): 8.

35. T. B. Harrington, "Factors Influencing Regeneration of Scotch Broom (*Cytisus scoparius*)," Pacific Northwest Research Station and USDA Forest Service, www.ruraltech.org/video/2006/invasive_plants/pdfs/NHS_Hall/17_harrington.pdf (accessed May 6, 2009).

36. Richo Cech, "Scotch Broom," www.horizonherbs.com (accessed December, 2008).

37. L. Gil, P. Fuentes-Utrilla, A. Soto, et al., "Phylogeography: English Elm is a 2,000-year-old Roman Clone," *Nature* 431, no. 7012 (October 28, 2004): 1053.

38. C. R. Hart, L. D. White, A. McDonald, et al., "Saltcedar Control and Water Salvage on the Pecos River, Texas, 1999–2003," *Journal of Environmental Management* 74, no. 4 (June 2005): 99–409.

39. Personal e-mail.

40. Hidegard von Bingen, *Hildegard von Bingen's Physica* (Rochester, Vt.: Healing Arts Press, 1998), 96.

41. H. W. Felter and John Uri Lloyd, *King's American Dispensatory* (Sandy, Ore.: Eclectic Medical Publications, 1898).

42. Nicholas Culpeper, *Complete Herbal.*

43. Personal e-mail.

Bibliography

Adams, M. "Cocktail of Pharmaceuticals Found in the Fish Caught Near Major U.S. Cities." March 26, 2009. www.naturalnews.com/025933.html. Accessed March 5, 2010.

Ausubel, Kenny, ed. *Ecological Medicine.* San Francisco: Sierra Club Books, 2004.

Barlow, Connie. *Ghosts of Evolution.* New York: Basic Books, 2000.

BASF Corporation documentary. "Plants out of Place 2." Invasive Weed Awareness Coalition.

Bateson, Gregory. *Steps to an Ecology of Mind.* New York: Ballantine Books, 1972.

Bensky, D., A. Gamble, S. Clavey, et al. *Chinese Herbal Medicine Materia Medica,* 2nd and 3rd editions. Seattle: Eastland Press, 1993, 2004.

Berry, Thomas. *The Dream of the Earth.* San Francisco: Sierra Club Books, 1988.

Beyfuss, Robert. "Evaluating the Invasive Potential of Imported Plants." *Journal of Medicinal Plant Conservation* (Winter 2007): 12–13.

Blumenthlal, D., R. A. Chimner, J. M. Welker, et al. "Increased Snow Facilitates Plant Invasion in Mixedgrass Prairie." *New Phytologist* 179, no. 2 (July 2008): 440–48.

Blumenthal M., W. R. Busse, A. Goldberg, et al., eds. *The Complete Commission E Monographs: Therapeutic Guide to Herbal Medicines.* Boston, Mass.: Integrative Medicine Communications, 1998.

Bohlen, Patrick J., Stefan Scheu, Cindy M. Hale, et al. "Non-native Invasive Earthworms as Agents of Change in Northern Temperate Forests." *Frontiers in Ecology and the Environment* 2, no. 8 (2004): 427–35.

Buhner, Stephen Harrod. *Healing Lyme: Natural Healing and Prevention for Lyme Borreliosis and Its Coinfections.* Randolph, Vt.: Raven Press, 2005.

———. *Herbal Antibiotics.* Pownal, Vt.: Storey Books, 1999.

———. *The Lost Language of Plants.* White River Junction, Vt.: Chelsea Green Publishing, 2002.

———. *The Secret Teachings of Plants.* Rochester, Vt.: Bear and Co., 2004.

Capra, Fritjof. *The Web of Life.* New York: Anchor Books, 1996.

Carson, Rachel. *Silent Spring.* New York: Houghton Mifflin, 1962.

Casey, Michael. "Removing Cats to Protect Birds Backfires on Island." Associated Press. January 13, 2009.

Caulcutt, C. "US: 'Superweed' Explosion Threatens Monsanto Heartlands." *France 24* news service. April 19, 2009.

Cech, Richo. *Making Plant Medicine.* Williams, Ore.: Horizon Herbs, 2000.

Chen, J., and T. Chen. *Chinese Medical Herbology and Pharmacology.* City of Industry, Calif.: Art of Medicine Press, 2004.

Chiej, Roberto. *The MacDonald Encyclopedia of Medicinal Plants.* London: MacDonald and Co., 1984.

Christopher, Tom. "Can Weeds Help Solve the Climate Crisis?" *New York Times.* June 29, 2008.

Cobbett, C. S. "Phytochelatin Biosynthesis and Function in Heavy-metal Detoxification." *Current Opinion in Plant Biology* 3, no. 3 (2000): 211–16.

Coffey, Timothy. *The History and Folklore of North American Wildflowers.* New York: Facts On File, 1993.

Comis, Don. "Phytoremediation: Using Plants To Clean Up Soils." *Agricultural Research* magazine (USDA) 48, no. 6 (June 2000). www.ars.usda.gov/is/AR/archive/jun00/soil0600.pdf. Accessed July 27, 2008.

Corn M. L., E. H. Buck, J. Rawson, et al. "Harmful Non-native Species: Issues for Congress." CRS Report for Congress, U.S. Congressional Research Service, Resources, Science, and Industry Division, Washington, D.C., April 8, 1999. www.ncseonline.org/NLE/CRSreports/Biodiversity/biodv-26.cfm. Accessed March 5, 2010.

Couplan, François. *The Encyclopedia of Edible Plants of North America.* New Canaan, Conn.: Keats Publishing, 1998.

Davis, M. A. "Biotic Globalization: Does Competition from Introduced Species Threaten Biodiversity?" *Bioscience* 53 (2003): 481–89.

Davis, M. A., K. Thompson, and J. P. Grime. "Charles S. Elton and the Dissociation of Invasion Ecology from the Rest of Ecology." *Diversity and Distributions* 7 (2001): 97–102.

Dazy, M., V. Jung, J. F. Férard, et al. "Ecological Recovery of Vegetation on a Coke-factory Soil: Role of Plant Antioxidant Enzymes and Possible Implications in Site Restoration." *Chemosphere* 74, no. 1 (December 2008): 57–63.

De Bairacli Levy, Juliette. *The Complete Handbook for Farm and Stable.* New York: Faber and Faber, 1991.

Derr, Mark. "Alien Species Often Fit In Fine, Some Scientists Contend." *New York Times.* September 4, 2001.

Devall, Bill, and George Sessions. *Deep Ecology.* Layton, Utah: Gibbs M. Smith, 1985.

Diamond, Jared. *Guns, Germs, and Steel.* New York: W. W. Norton, 1999.

Donn, Jeff, Martha Mendoza, and Justin Pritchard. "AP IMPACT: Tons of Released Drugs Taint US Water." Associated Press, April 19, 2009.

Dov, F. Sax, and Steven D. Gaines. "Colloquium Paper: Species Invasions and Extinction: The Future of Native Biodiversity on Islands." *Proceedings of the National Academy of Sciences* 105, Suppl. 1 (2008): 11490–97.

Duke, J. A. *Handbook of Medicinal Herbs.* Boca Raton, Fla.: CRC Press, 1985.

Duke. J. A., and E. S. Ayensu. *Medicinal Plants of China,* vols. 1 and 2. Algonac, Mich.: Reference Publications, 1985.

Duke, S. O., A. C. Blair, F. E. Dayan, et al. "Is (-)-catechin a Novel Weapon of Spotted Knapweed *(Centaurea stoebe)*?" *Journal of Chemical Ecology* 35, no. 2 (February 2009): 141–53.

Elton, Charles S. *The Ecology of Invasions by Animals and Plants.* Chicago: University of Chicago Press, 1958.

"Exposure to Rainbow Herbicides." www.agent-orange-lawsuit.com/index.html; www.allmilitary.com/board/viewtopic.php?id=1209. Accessed March 5, 2010.

Foster, Stephen, and Christopher Hobbs. *A Field Guide to Western Medicinal Plants and Herbs.* New York: Houghton Mifflin, 2002.

Foster, Stephen, and James A. Duke. *A Field Guide to Medicinal Plants.* New York: Houghton Mifflin, 1990.

H. W. Felter, and John Uri Lloyd. *King's American Dispensatory,* vols. 1 and 2. Sandy, Ore.: Eclectic Medical Publications, 1898.

Frank, Joseph. *The Invasive Species Cookbook.* Wauwatosa, Wis.: Bradford Street Press, 2007.

Fukuoka, Masanobu. *The Natural Way of Farming.* Madras, India: Bookventure, 1985.

———. *The One-Straw Revolution.* Emmaus, Pa.: Rodale Press, 1978.

Gattuso, Dana Joel. "Invasive Species: Animal, Vegetable or Political?" *National Policy Analysis* 544, August 2006. www.nationalcenter.org/NPA544InvasiveSpecies .html. Accessed March 5, 2010.

Gaynor, Tim, and Steve Gorman. "Fast-growing Western U.S. Cities Face Water Crisis." Reuters. March 11, 2009.

Gillam, Carey. "Biotech Crops Cause Big Jump in Pesticide Use: Report." Rueters. Nov 17, 2009.

Gladstar, Rosemary, and Pamela Hirsch, eds. *Planting the Future.* Rochester, Vt.: Healing Arts Press, 2000.

"Goats Provide a Non-Toxic Alternative to Spraying Herbicides for Invasive Weeds." *Beyond Pesticides.* May 10, 2002. www.beyondpesticides.org/news/daily_news_ archive/2002/05_10_02.htm. Accessed March 5, 2010.

Gómez Álvarez-Arenas, T. E., D. Sancho-Knapik, J. J. Peguero-Pina y, and E. Gil-

Pelegrín. "Noncontact and Noninvasive Study of Plant Leaves Using Air-coupled Ultrasounds." *Applied Physics Letters* 95, no. 19 (2009): 193702.

Green, James. *The Herbal Medicine Maker's Handbook.* Berkeley, Calif.: The Crossing Press, 2000.

Grieve, Maud. *A Modern Herbal,* vols. 1 and 2. New York: Dover Publications, 1971.

Gruenwald J., T. Brendler, C. Jaenicke, eds. *PDR for Herbal Medicines,* 2nd edition. Montvale, N.J.: Medical Economics Co., 2000.

Gurevitch, Jessica, and D. K. Padilla. "Are Invasive Species a Major Cause of Extinctions?" *Trends in Ecology and Evolution* 19, no. 9 (September 2004): 470–74.

Gutierrez, David. "Hospitals Flush 250 Million Pounds of Expired Drugs Into Public Sewers Every Year." *Natural News.* February 10, 2009. www.naturalnews .com/025573_drugs_water_hospitals.html. Accessed March 5, 2010.

———. "Weed Killer Chemicals Linked to Brain Cancer." *Natural News.* October 8, 2008.

Haas, Elson M. *Staying Healthy with Nutrition.* Berkeley: Celestial Arts, 2006.

Harris, Ben C. *Eat the Weeds.* New Canaan, Conn.: Keats Publishing, 1969.

Hemenway, Toby. "Native Plants: Restoring to an Idea." www.patternliteracy.com/ nativeplantsres.html. Accessed March 5, 2010.

———. *Gaia's Garden.* White River Junction, Vt.: Chelsea Green Publishing, 2000.

Henry, Jeanna R. "An Overview of the Phytoremediation of Lead and Mercury." National Network of Environmental Management Studies (NNEMS) status report, 2000. www.kossge.or.kr/data/pdf/an%20Overview%20of%20the%20 phytoremediation%20Pb%20Hg.pdf. Accessed April 16, 2009.

Hierro J. L., D. Villarreal, O. Eren, et al. "Disturbance Facilitates Invasion: The Effects Are Stronger Abroad Than at Home." *The American Naturalist* 169, no. 2 (August 2006): 144–56.

Hoffmann, David. *Medical Herbalism.* Rochester, Vt.: Healing Arts Press, 2003.

———. *The Herbal Handbook.* Rochester, Vt.: Healing Arts Press, 1987.

Holmes, Peter. *The Energetics of Western Herbs,* vols. 1 and 2. Berkeley: NatTrop Publishing, 1993.

Huntley, Chris, ed. "Dow AgroSciences Unveils Breakthrough in Multiple Herbicide Tolerance Traits." Dow Chemical Company news release, http://news.dow.com/ dow_news/corporate/2007/20070828a.htm. Accessed March 5, 2010.

Jacke, D., and E. Toensmeier. *Edible Forest Gardens,* vols. 1 and 2. White River Junction, Vt.: Chelsea Green Publishing, 2005.

Johnson, Tim. "Invasive Species." *Burlington Free Press.* November 9, 2003, 1D.

———. "UN: Killer Strains of Tuberculosis May 'Spiral out of Control.'" *McClatchy Newspapers,* Apr 1, 2009.

Kaufman, Wallace. "Invasion of Alien Species." *PERC Reports* 21, no. 4 (December 2003).

Kaufman, Sylvan Ramsey, and Wallace Kaufman. *Invasive Plants: A Guide to Identification, Impacts, and Control of Common North American Species.* Mechanicsburg, Pa.: Stackpole Books, 2007.

Khan, A. G., C. Kuek, T. M. Chaudhry, et al. "Role of Plants, Mycorrhizae and Phytochelators in Heavy Metal Contaminated Land Remediation." *Chemosphere* 21 (2000): 197–207.

Khare, C. P. *Indian Herbal Remedies.* New York: Springer, 2004.

Liu, J. "Pharmacology of Oleanolic Acid and Ursolic Acid." *Journal of Ethnopharmacology* 49, no. 2 (December 1, 1995): 57–68.

LiveScience Staff. "Pets Pass Superbug to Humans." *LiveScience,* June 21, 2009. www.livescience.com/health/090621-mrsa-dogs.html. Accessed June 22, 2009.

Lockwood, Julie. *Invasion Ecology.* Malden, Mass.: Blackwell Publishing, 2007.

Lodge, David M., and Kristin Shrader-Frechette. "Nonindigenous Species: Ecological Explanation, Environmental Ethics, and Public Policy." *Conservation Biology* 17 (2003), 31–37.

Louda, S. M., D. Kendall, J. Connor, et al. "Ecological Effects of an Insect Introduced for the Biological Control of Weeds." *Science* 277, no. 5329 (August 22, 1997): 1088–90.

Lovelock, James. *Healing Gaia.* New York: Harmony Books, 1991.

Mabberley, David J. *Mabberley's Plant-Book,* 3rd edition. Cambridge, England: Cambridge University Press, 2008.

MacDougall, A. "Did Native Americans Influence the Northward Migration of Plants During the Holocene?" *Journal of Biogeography,* 30, no. 5 (May 2003): 633–47.

Marinelli, J, and J. M. Randall, eds. *Invasive Plants (Weeds of the Global Garden).* Brooklyn, N.Y.: Brooklyn Botanical Garden, 1996.

Mason, Margie, and Martha Mendoza. "First Case of Highly Drug-resistant TB Found in US." Associated Press, December 27, 2009.

Mellon, Margaret, and Jane Rissler. "Environmental Effects of Genetically Modified Food Crops—Recent Experiences." Union of Concerned Scientists. June 12–13, 2003.

Mendoza, Martha. "Study: Range of Pharmaceuticals in Fish across US." Associated Press, March 25, 2009.

Mendoza, Martha, and Margie Mason. "Solution to Killer Superbug Found in Norway." Associated Press, December 31, 2009.

Millar, Constance I., and Linda B. Brubaker. "Climate Change and Paleoecology: New Contexts for Restoration Ecology." In M. Palmer, D. Falk, and J. Zedler, eds. *Restoration Science.* Washington, D.C.: Island Press, 2006, 315–40.

Moffat, A. S. "Global Nitrogen Overload Problem Grows Critical." *Science* 279, no. 5353 (February 13, 1998): 988–89.

Monsanto News Release. "Monsanto Revises Full-Year 2008 Earnings Per Share Guidance Based on Better Than Anticipated Results for Seeds and Traits." September 16, 2008. PRNewswire-FirstCall.

Mooney, H. A., and E. E. Cleland. "The Evolutionary Impact of Invasive Species." *PNAS* 98, no. 10 (2001): 5446–451.

Moore, Michael. *Medicinal Plants of the Desert and Canyon West.* Santa Fe, N.M.: Museum of New Mexico Press, 1989.

———. *Medicinal Plants of the Mountain West.* Santa Fe, N.M.: Museum of New Mexico Press, 1979.

Naidenko O, N. Leiba, R. Sharp, et al. "Bottled Water Contains Disinfection Byproducts, Fertilizer Residue, and Pain Medication." *Environmental Working Group.* October 2008.

National Aeronautics and Space Administration (NASA). "Solar Variability: Striking A Balance with Climate Change." *ScienceDaily.* May 12, 2008. www.sciencedaily.com/releases/2008/05/080512120523.htm. Accessed June 6, 2009.

National Aeronautics and Space Administration (NASA). "Success at Stennis—A New Image for the Water Hyacinth." http://technology.ssc.nasa.gov/suc_stennis_water.html. Accessed November 23, 2008.

National Agricultural Library (USDA). "Laws and Regulations of Invasive Species." www.invasivespeciesinfo.gov/laws/execorder.shtml. Accessed February 5, 2008.

National Wildlife Refuge Association. "Silent Invasion: A Call to Action from NWRA." October 2002. www.refugenet.org/new-pdf-files/Silent%20Invasion%20pdf.pdf. Accessed March 5, 2010.

Nature Conservancy. "Americas Least Wanted." *The Nature Conservancy* (1996). www.tnc.org.

Nuzzo, V. A., J. C. Maerz, and B. Blossey. "Earthworm Invasion as the Driving Force Behind Plant Invasion and Community Change in Northeastern North American Forests." *Conservation Biology.* February 18, 2009.

Ottenhof, C. J., A. Faz Cano, J. M. Arocena, et al. "Soil Organic Matter from Pioneer Species and Its Implications to Phytostabilization of Mined Sites in the Sierra de Cartagena (Spain)." *Chemosphere* 69, no. 9 (November 2007): 1341–50.

Parmesan, C. "Ecological and Evolutionary Responses to Recent Climate Change." *Annual Review of Ecology and Systematics* 37 (2006): 637–69.

Pearson D. E. "Invasive Plant Architecture Alters Trophic Interactions by Changing Predator Abundance and Behavior." *Oecologia* 159, no. 3 (March 2009): 549–58.

Pendell, Dale. *Living with Barbarians.* Sebastopol, Calif.: Wild Ginger Press, 1999.

———. *Pharmako/poeia.* San Francisco: Mercury House, 1995.

Perry, L. G., G. C. Thelen, W. M. Ridenour, et al. "Concentrations of the Allelochemical Catechin in *Centaurea maculosa* Soils." *Journal of Chemical Ecology* 33 (2007): 2337–44.

Peterson, Lee Allen. *A Field Guide to Edible Wild Plants*. New York: Houghton Mifflin, 1977.

Pfeiffer, Ehrenfried. *Weeds and What They Tell*. Emmaus, Pa.: Rodale Press, n.d.

Pimetel, David, and Christa Wilson. "Economic and Environmental Benefits of Biodiversity." *BioScience,* no. 11 (December, 1997): 747–57.

Pimentel, D., L. Lach, R. Zuniga, et al. "Environmental and Economic Costs of Non-indigenous Species in the United States." *BioScience* 50 (2000): 53–65.

Plataforma SINC. "Native Plants Can Also Benefit From The Invasive Ones." *ScienceDaily*. May 21, 2008. www.sciencedaily.com/releases/2008/05/080516125934.htm. Accessed April 15, 2009.

Rabinovitch, Ari. "Scientists 'Listen' to Plants to Find Water Pollution." Reuters. August 14, 2008.

Raskin, Ilya. *Phytoremediation of Toxic Metals: Using Plants to Clean Up the Environment*. New York: John Wiley and Sons, 2000.

Reid, Daniel. *The Tao of Detox*. Rochester, Vt.: Healing Arts Press, 2006.

Reinberg, Steven. "Study Links Agent Orange to Prostate Cancer in Vietnam Vets." *Washington Post*. August 6, 2008.

Revolutionary Health Committee of Hunan Province. *A Barefoot Doctors Manual*. Seattle: Cloudburst Press, 1977.

Roach, John. "'Grass Gas' Shows Promise as Superefficient, Clean Fuel." *National Geographic News*. January 9, 2008.

Roberson, Emily B. "Plant Science and Conservation Groups Ask Congress to Add Plants to Legislation Protecting Wildlife from Climate Change." Native Plant Conservation Campaign, September 18, 2007. www.plantsocieties.org/Equal_Protection.htm. Accessed March 5, 2010.

"Roundup Herbicide Toxicity." www.naturescountrystore.com/roundup/index.html. Accessed March 5, 2010.

Sagoff, Mark. "Do Invasive Species Threaten the Environment?" *Journal of Agricultural and Environmental Ethics* 18 (2005): 215–36. www.propertyrightsresearch.org/2005/articles04/do_invasive_species_threaten_the.htm. Accessed March 5, 2010.

Schmid, Randolph E. "Ants, Plants Mutually Benefit Each Other." Associated Press. January 10, 2008.

Schmitz, D. C., and D. Simberloff. "Biological Invasions: A Growing Threat." *Issues in Science and Technology* 13 (Summer 1997): 33–41. www.nap.edu/issues/13.4/schmit.htm. Accessed March 5, 2010.

Schuster W. S., K. L. Griffin, H. Roth, et al. "Changes in Composition, Structure and Aboveground Biomass over Seventy-six Years (1930–2006) in the Black Rock Forest, Hudson Highlands, Southeastern New York State." *Tree Physiology* 28, no. 4 (April 2008): 537–49.

Schwab, C. "Chemicals: The Good, the Bad, and the Deadly—the Mystery Club." National Ag Safety Database (a division of the CDC). www.nasdonline.org/docs/ d001801-d001900/d001888/d001888.html. Accessed March 5, 2010.

Science Encyclopedia (jrank.org). "Chemical Warfare—Use of Herbicides During The Vietnam War." http://science.jrank.org/pages/1391/Chemical-Warfare-Use-herbicides-during-Vietnam-War.html. Accessed March 5, 2010.

Smith, Betty. *A Tree Grows in Brooklyn.* New York: HarperCollins, 1943.

Society for General Microbiology. "Spreading Antibiotics in the Soil Affects Microbial Ecosystems." *ScienceDaily.* April 2, 2009. www.sciencedaily.com/ releases/2009/03/090329205445.htm. Accessed April 22, 2009.

Spelman, Kevin. "'Silver Bullet' Drugs vs. Traditional Herbal Remedies: Perspectives on Malaria." *HerbalGram* 84 (Nov 2009): 45–55.

Stamets, Paul. *Mycelium Running: How Mushrooms Can Help Save the World.* Berkeley: Ten Speed Press, 2005.

Stamati F. E., N. Chalkias, D. Moraetis, and N. P. Nikolaidis. "Natural Attenuation of Nutrients in a Mediterranean Drainage Canal." *J Environ Monit* 12, no. 1 (2010): 164–71.

Starfield B. "Is US Health Really the Best in the World?" *JAMA* 284, no. 4 (July 26, 2000): 483–85.

Streater, Scott. "US: Saharan Dust May Have You Wheezing." *Star-Telegram.* June 25, 2008.

Tharayil, N, P. Bhowmik, P. Alpert, et al. "Dual Purpose Secondary Compounds: Phytotoxin of *Centaurea diffusa* also Facilitates Nutrient Uptake." *The New Phytologist* 181, no. 2 (January 2009): 424–34.

Theodoropoulos, David. *Invasion Biology: Critique of a Pseudoscience.* Blythe, Calif.: Avvar Books, 2003.

Tierra, Michael. *The Way of Herbs.* New York: Pocket Books, 1998.

Tilford, Gregory L. *Edible and Medicinal Plants of the West.* Missoula, Mont.: Mountain Press Publishing, 1997.

Tokar, B. "Agribusiness, Biotechnology and War." Institute for Social Ecology Biotechnology Project. September 1, 2002. www.social-ecology.org/2002/09/ agribusiness-biotechnology-and-war. Accessed March 5, 2010.

Tompkins, Peter, and Christopher Bird. *Secrets of the Soil.* Anchorage, Alaska: Earthpulse Press, 1998.

———. *The Secret Life of Plants.* New York: Avon Books, 1973.

Trefil, James. "When Plants Migrate—The Study of How Plants Moved North after the Last Ice Age Could Mean New Directions for Conservation." *Smithsonian.* September 1998.

United States Department of Agriculture. USDA, NRCS. 2010. The PLANTS Database (http://plants.usda.gov). National Plant Data Center, Baton Rouge, La 70874-4490 USA.

USDA, ARS, National Genetic Resources Program. Germplasm Resources Information Network (GRIN). National Germplasm Resources Laboratory, Beltsville, Maryland. www.ars-grin.gov/cgi-bin/npgs/html/noxweed.pl.

United States Department of Interior, U.S. Geological Survey, National Invasive Species Information Center. http://wfrc.usgs.gov/research/invasiveintro.htm.

United States Geological Survey. www.USGS.gov.

United States National Arboretum. www.USNA.USDA.gov.

University of California, Santa Barbara. "Is Extinction or Diversity on the Rise? Study of Islands Reveals Surprising Results." *ScienceDaily.* August 27, 2008. www.sciencedaily.com /releases/2008/08/080826173227.htm. Accessed August 27, 2008.

Uphof, J. C. *The Dictionary of Economic Plants.* New York: Stechert-Hafner Service Agency, 1968.

Verkleij, J. A. C., F. Sneller, and H. Schat. "Metallothioneins and Phytochelatins: Ecophysiological Aspects." In Y. P. Abrol and A. Ahmad, eds. *Sulphur in Plants.* The Netherlands: Kluwer Academic Publishers, 2003.

Vermeij, G. "An Agenda for Invasion Biology." *Biological Conservation* 7 (1996): 83–89.

———. "When Biotas Meet: Understanding Biotic Interchange." *Science* 253, no. 5024 (September 6, 1991).

Vollmer, Sabine. "US: Pesticide Data May Tell Why Bees Die." *News Observer.* August 22, 2008.

von Bingen, Hidegard. *Hildegard von Bingen's Physica.* Rochester, Vt.: Healing Arts Press, 1998.

Weed Science Society of America. "Weed Science Society of America Warns Glyphosate Resistance Increasing." www.wssa.net/WSSA/PressRoom/WSSA_Glyphosate_Resistance.pdf. Accessed March 5, 2010.

Wessels, Tom. *Reading the Forested Landscape.* Woodstock, Vt.: The Countryman Press, 1997.

Yarnell, E, and K. Abascal. "Dilemmas of Traditional Botanical Research." *HerbalGram* 55 (2002): 46–54.

Yeung, Him-che. *Handbook of Chinese Herbs,* 2nd edition. Rosemead, Calif.: Institute of Chinese Medicine, 1996.

Zimmer, Carl. "Friendly Invaders." *New York Times.* September 8, 2008.

Ziska, L. H. "Evaluation of the Growth Response of Six Invasive Species to Past, Present and Future Atmospheric Carbon Dioxide." *Journal of Experimental Botany* 54 (2003): 395–404.

Index

Bond cycles, 33
brake fern (*Pteris vittata*), 104
Brassica rapa (wild mustard), pl.23, 308–13
Buhner, Stephen, 122, 131, 141, 178, 223
Burbank, Luther, 119–20
burdock, 155

cadmium, 154
caffeic acid, 162–63
California, 33
California Exotic Pest and Plant
    Council, 11
California Farm Bureau, 14
Canadian waterweed, 23
canals, 47–48
cancer, 151
cane toads, 65
Carson, Rachel, 67
Carver, George Washington, 119–20
catechin, 62, 160–61
cattails, 56
C. calcitrapa (red star thistle), 233
C. cyanus (cornflower), 233
Cech, Richo, 173
Celastrus orbiculatus (Oriental
    bittersweet), pl.11, 243–47
Celtis spp. (hackberry), 53
Centaurea diffusa (diffuse knapweed), 62
Centaurea maculosa (spotted knapweed),
    pl.10, 62, 231
Centaurea spp. (knapweed), 109, 230–36
Center for Disease Control (CDC), 139,
    152–53
Chaney, Rufus, 85, 104–5
chelation, 154–65
China, 41
cholera, 146–47
Christopher, Tom, 37, 58
C. iberica (Iberian star thistle), 231
Cicuta maculata (water hemlock), 64

cilantro, 155
Cirsium spp. (thistle), pl.19–20, 289–94
climate change, 24, 33–39, 35, 60
Clinton, William, 10
collection of plants, 172. *See also* specific
    plants
communication of plants, 119–25
Congress, U.S., 10, 13
Conservation Biology, 48–49
consumption, 144
Convolvulus arvensis (bindweed), 191–95
copper, 153
corn ethanol, 82–83
cornflower (*C. cyanus*), 233
cow parsnip (*Heracleum lanatum*), 64
Cryptosporidium, 146–47
C. solstitialis (yellow star thistle), 231,
    233–34, 235
Cytisus scoparius (Scotch broom), pl.15,
    62, 273–77

Dalton, David, 141
dams, 47–48
dandelion (*Taraxacum officinale*), 62,
    109, 155, 201–7
Davis, M. A., 24
decoctions, 177
deep ecology, 92–95
    health and, 95–96
    phytoremediation, 100–115
    pollution and, 96–100
deer ticks, 139–40
dengue fever, 142–44
development, 55
Dias, Michael, 14
diffuse knapweed (*Centaurea diffusa*), 62
diseases, 130–32
    antibiotics and, 132–33
    chart of plants and pathogens, 148–49
    current status of, 133–36

BOOKS OF RELATED INTEREST

The Secret Teachings of Plants
The Intelligence of the Heart in the Direct Perception of Nature
by Stephen Harrod Buhner

Sacred Plant Medicine
The Wisdom in Native American Herbalism
by Stephen Harrod Buhner

Plant Spirit Healing
A Guide to Working with Plant Consciousness
by Pam Montgomery

Plant Spirit Shamanism
Traditional Techniques for Healing the Soul
by Ross Heaven and Howard G. Charing

Morphic Resonance
The Nature of Formative Causation
by Rupert Sheldrake

The Presence of the Past
Morphic Resonance and the Habits of Nature
by Rupert Sheldrake

The Rebirth of Nature
The Greening of Science and God
by Rupert Sheldrake

Alchemical Medicine for the 21st Century
Spagyrics for Detox, Healing, and Longevity
by Clare Goodrick-Clarke

INNER TRADITIONS • BEAR & COMPANY
P.O. Box 388
Rochester, VT 05767
1-800-246-8648
www.InnerTraditions.com

Or contact your local bookseller